A GENERAL RELATIVITY COURSEBOOK

General relativity is a subject that most undergraduates in physics are particularly curious about, but it has a reputation for being very difficult. This book provides as gentle an introduction to general relativity as possible, leading you through the necessary mathematics in order to arrive at important results. Of course, you cannot avoid the mathematics of general relativity altogether, but, using this book, you can gain access to and appreciation of tensors and differential geometry at a pace you can keep up with. Early chapters build up to a complete derivation of Einstein's equations, while the final chapters cover the key applications on black holes, cosmology and gravitational waves. It is designed as a coursebook with just enough material to cover in a one-semester undergraduate class, but it is also accessible to any numerate readers who wish to appreciate the power and beauty of Einstein's creation for themselves.

ED DAW is Professor of Particle Astrophysics at the University of Sheffield. He has worked as an experimental physicist since 1998, on searches for dark matter and gravitational waves. His work on gravity led him to volunteer to teach general relativity at Sheffield, which he has continued to do from 2003 until the present. He considers general relativity a hobby, albeit one that is crucial to underpin his understanding of his own research. He also enjoys trying to explain hard things in simple terms, a very good habit for a professor.

The approach in the book is unique, and especially valuable for the student first encountering general relativity. It shows in detail the computational steps involved in gaining the main results.

RAINER WEISS, Nobel laureate; Professor Emeritus at MIT

I think this is an excellent introduction to general relativity, and its important applications to cosmology and gravitational wave astrophysics, for the serious student who has not experienced the necessary mathematical formalism before and is willing to follow the text and attempt the many examples. It is an ideal lead-in to many of the more sophisticated modern textbooks which are now available.

SIR JAMES HOUGH, OBE FRS FRSE; University of Glasgow

A GENERAL RELATIVITY COURSEBOOK

ED DAW

University of Sheffield

CAMBRIDGE
UNIVERSITY PRESS

Shaftesbury Road, Cambridge CB2 8EA, United Kingdom

One Liberty Plaza, 20th Floor, New York, NY 10006, USA

477 Williamstown Road, Port Melbourne, VIC 3207, Australia

314–321, 3rd Floor, Plot 3, Splendor Forum, Jasola District Centre,
New Delhi – 110025, India

103 Penang Road, #05-06/07, Visioncrest Commercial, Singapore 238467

Cambridge University Press is part of Cambridge University Press & Assessment,
a department of the University of Cambridge.

We share the University's mission to contribute to society through the pursuit of
education, learning and research at the highest international levels of excellence.

www.cambridge.org
Information on this title: www.cambridge.org/highereducation/isbn/9781009242431

DOI: 10.1017/9781009242479

First published 2023

A catalogue record for this publication is available from the British Library.

A Cataloging-in-Publication data record for this book is available from the Library of Congress

ISBN 978-1-009-24243-1 Hardback
ISBN 978-1-009-24244-8 Paperback

Additional resources for this publication at www.cambridge.org/daw-GR

Dedicated to my father, Stephen Francis Daw, 1944–2012

Contents

Preface

General relativity (GR) is one of the most fascinating areas of physics. It is therefore naturally an attractor for undergraduates, often given as a reason for choosing the subject. It is not, however, always offered as an undergraduate course because it is hard to teach. I have taught an undergraduate GR course at Sheffield since 2004 (with a few years off at one point). I have developed some strategies that have allowed the course to continue to be successful. My notes are now mature enough that I offer them up as a coursebook, hoping it will be of service to others faced with the same task.

I try to adhere to the maxim of uncovering some of the material rather can covering all of it. A detailed description of the tensor formalism is unavoidable. I use what some would consider an old fashioned approach, defining tensors in terms of the transformation properties of their components. The book contains all the mathematical detail students need to arrive at important results. In this sense, it differs markedly from other books, which tend to leave the lengthier components of derivations as stated without proof or for the students to derive. Experience has taught me that students are simply not ready for such long tracts of algebra, or they don't have the time for it. So, almost everything is there, even though the algebra is sometimes tedious. I also leave in all the factors of c and G, so the whole book is in SI (MKS) units. These constants also allow you to maintain constant contact with the experimental world, to assess the relative magnitudes of terms, and to check your algebra with dimensional analysis.

There are three chapters on applications, on the Schwarzschild solution, the Friedman–Lemaître–Robertson–Walker cosmology, and gravitational waves. The chapters contain the GR at the core of these areas; the problems at the end contain selected applications. Many of the problems are therefore foundational general relativity. I find it pedagogically far better to have the students go through this work for themselves than to teach it as bookwork, as in my experience students fail to digest the latter and it is almost immediately forgotten. The problems are at a spectrum of levels of difficulty from extremely easy to quite long and involved. The idea is that there is something there for students having a wide range of abilities and levels of preparedness. There is also an example of what I refer to as a 'diagnostic quiz' in Section 1.8. I set a quiz like this to students starting the course well before the drop date, as a refresher course and a friendly warning shot across the bows for those who have embarked upon the course not understanding the required level of mathematical competency. I recommend this practice to my fellow teachers.

I find that I can cover the material up to the end of Section 8.8 in 18–20 50 minute lectures and that students who spend perhaps 6 hours a week studying for themselves succeed in learning the material. If you only have 18 lectures, then you can move quite quickly through the calculations leading to the Einstein tensors in Chapters 6 and 7, though your course will be improved if you can dwell in these areas and give students a strong feel for the weight and work represented in these calculations. At the end of the book, starting in Section 8.9.1, I have included some 'optional extras'. Firstly, there is a look at some of the technical aspects of practical experimental gravitational wave interferometry. This is not general relativity, but I believe that the subject should not live in isolation from the rest of science. It can be by all means left out of a one semester course. The final chapter is a guide to further reading, a summary of conventions in differential geometry that differ from mine, and a short review of the more advanced literature in various research fields to which general relativity is connected.

I do not know if I have succeeded in walking the balance beam between the abyss of a book that is too advanced and the disappointment a book that is too superficial. Now that I consider the book in its entirety, I see that it might be of interest beyond the original target audience. Time will tell. I hope at least to convey the love of the subject that I have gained through familiarity and that you enjoy what is between these covers.

Acknowledgments

When my colleagues at Sheffield found out that I had written this book, many of them asked how I had found the time. The book evolved as I addressed areas of confusion that I perceived amongst the students. So I thank the students for their many questions, comments and corrections. I am sure that many times at home I have been staring at my computer or into space when I should have been with my family. I thank my wife Anne and my children Georgia and Eli for their patience, encouragement and love during this journey. I would also like to thank Pieter Kok for a careful reading and my colleagues at LIGO for numerous invigorating discussions, which informed my thinking in many areas, most significantly for Chapter 8.

1

The Principle of Equivalence

1.1 Are Smooth Curves Natural?

When you first encounter calculus, the very first thing you are usually taught to do is to measure the gradient of a curve. Usually, the functional form of that curve is some polynomial, for example, $y = 3x^2$. You are shown how to multiply the coefficient of x^2 by the power and then decrease the power by one unit, so that the slope in this case is $dy/dx = 6x^1$, and probably asked to evaluate the slope at some value of x, for example, the slope is 12 where $x = 2$. This machinery was invented by Newton (Newton et al. 1999), although others, notably Leibniz (Roy 2021), had similar ideas around the same time, and there was closely related work outside the west, for example, at the Kerala school in India (Katz 1995), that predated Newton's renowned contributions.

I think it is fair to say that most introductory calculus courses spend little time worrying about the soundness of the assumptions underlying these mathematical methods. That is because you meet calculus first in mathematics and there are lots of abstract functions for which those assumptions seem to be safe. As physicists, however, our job is to draw parallels between the behaviour of nature, which we measure, and the properties of abstract mathematics. Does nature adhere to the fundamental assumptions underlying calculus?

So, let us look at the central assumption that a curve is differentiable. What does this mean? It means that if I observe a portion of a curve with ever-greater magnification, then the ever-smaller sections of the curve look straighter and straighter. If this is true, then calculus will work, because it gets easier and easier to estimate the slope of a curve as it looks straighter and straighter, and the estimates of the slope based on examining ever-smaller portions of a curve agree with each other better and better as the magnification increases. This means that the concept of a limit is useful – the gradient of a curve is the estimated slope based on extrapolation of the estimates of the slope based on looking at portions of the curve where the magnification tends to infinity. Mathematically, the gradient of a function $y(x)$ of one variable x is

$$\frac{dy}{dx} = \lim_{\varepsilon \to 0} \frac{y(x + \varepsilon) - y(x)}{\varepsilon}. \tag{1.1}$$

You might think that this is fine for abstract polynomials, although the idea of the ratio of two quantities that are both approaching zero being finite and well behaved is not without its problems, and mathematicians have had to place stronger foundations under the concept of a derivative. Putting these concerns aside, is the concept of a derivative valid in nature? Are nature's curves differentiable?

Some are not. Material objects are ultimately made up of atoms, which do not lend themselves to the formation of mathematical smooth curves. A metal straight-edge from a machine shop certainly looks like a mathematical straight line. However, under a powerful microscope our 'straight-edge' actually looks less and less smooth as the magnification increases. Although this might worry us from a philosophical standpoint, the straight-edge is still very useful for checking the flatness and smoothness of pieces you are fabricating in a machine shop. This works because when you engineer something, you are working to some agreed tolerance, say a micrometer. As long as your straight edge is flat to within one micrometre, it is good enough to check the straightness of other macroscopic objects, also to within similar tolerances. So, our mathematical abstract concept of straightness is bought into approximate correspondence with the realities of a machine shop by agreeing that we will not worry about the breakdown of nature's adherence to the mathematical idea of straightness, as long as the departures between nature and mathematics occur at distance scales that are less than a micrometre. A micrometre is, as a scientist might say, microscopically large but macroscopically small. We can also machine curved objects that look like mathematical curves on a distance scale above a micron, although again when examined more closely, the resemblance disappears.

You might also ponder other curves in nature. When I throw a ball into the air, it follows a curve because of the pull of gravity on the ball. You can imagine this curve in your minds eye, and making the assumption that the acceleration due to gravity close to Earth's surface is of constant magnitude g directed downwards, the form of the curve is

$$\begin{pmatrix} x(t) \\ y(t) \end{pmatrix} = \begin{pmatrix} u^x t \\ u^y t - \frac{g}{2} t^2 \end{pmatrix}, \tag{1.2}$$

where u^x and u^y are the components of the velocity with which the ball was thrown parallel to and perpendicular to the ground, and t is time. This curve is a parabola from the abstract world of mathematics. How good a model is it for the actual path of the ball? Actually, it is very good but not perfect. For a start, there is air resistance and turbulence and wind, and of course the acceleration due to gravity is not exactly the same everywhere, then there is the rotation of the Earth, so that the observer standing on the ground is not actually at rest or even moving at a constant velocity, so that the ball is subject to pseudoforces in the accelerating frame of reference of the person who threw it. We could take away many of these approximations in theory by repeating the experiment in an evacuated enclosure, and we could use a mathematical model that accounted for pseudoforces and model the Earth's gravitational field more carefully in the evacuated box. We also have to note that the ball is of finite extent, so that it is the centre of mass of the ball that is meant to follow a smooth mathematical curve. To observe the ball, it has to be under bombardment by the quanta that make up light, which impart some momentum to the ball and introduce random fluctuations to the pathway that the ball actually takes about the smooth abstract curve. Finally, there is the idea that the very nature of the matter making up the ball is due to its coupling to the Higgs boson, so that anything moving at a velocity less than that of light is retarded to its pedestrian pace by constant interactions with Higgs bosons pulled out of the vacuum of standard model particle physics. You can see that the more troubling of these effects involve the underlying quantum nature of reality. Again, at small enough scales in distance or at high enough precision, nature appears to behave in a way that is very hard to model with calculus.

1.2 The Birth of General Relativity

General relativity was invented before quantum mechanics, before we had realised the deep ramifications of the microscopic world and its strange properties. This was actually a great blessing if it meant that Einstein was not overly distracted by worries about whether the smooth geometric world he imagined corresponded to reality. He could proceed to work out the consequences of an idea that he had, which was actually quite simple. The idea can be uncovered by going back to our ball flying through the air. We notice that in Equation (1.2), the mass m of the ball does not appear. This led Einstein to think of the pathway taken by an object freely falling in a gravitational field as being a property of space through which the object is moving. You could take objects of different masses and throw them all into the air with the same initial velocity, and according to Equation (1.2), their centres of mass would all follow the same path. This path is curved, as we can all see when we watch a ball trace it out. Therefore, in some sense, the presence of the large mass of the Earth is causing space and time to become locally curved. This was nicely summarised by John Wheeler:

Space-time tells matter how to move; matter tells space-time how to curve

(John Archibald Wheeler, 2000)

If we were only interested in ethereal ideas, this fascinating statement might be all we needed. The curving of space-time causes bodies in free fall to follow curved arcs, called geodesics. Equally, the curvature of those geodesics is due to the presence of massive bodies. The two concepts are actually equivalent. Curvature of space-time arises due to matter, matter gives rise to space-time curvature.

However, we are physicists, so we must learn the mathematics that give precise meaning to these statements. What is a curved space, which we know from special relativity is part of a broader concept called space-time? How do you quantify the presence of mass, which we know from special relativity is also the presence of energy? The definition of curvature will be made using the machinery of calculus, which when conducted in higher dimensions than two is called differential geometry. In using calculus to describe the curvature of the space-time of nature, we are assuming that space-time can be considered smooth at all. We know from the earlier discussion that this is a bold assumption.

Ultimately, the smooth space-time that general relativity requires must interface with the microscopic world. Perhaps this indicates that general relativity is an effective theory based on some underlying more fundamental formalism that is unified with the quantum field theory of the standard model. Or, perhaps, gravity is an emergent phenomenon that appears at distance scales that are microscopically large. The smoothness necessary to model it with general relativity could be a sufficiently good approximation at the scales where gravity is observed. At microscopic scales where the smoothness of space-time might cease to be valid, gravity could cease to be a force of nature. After all, it is famously difficult to study gravity at small distance scales; current best efforts probe gravity down to a distance scale of approximately 10 micrometers, a similar scale to the tolerances to which high-quality machine shop straight edges are made. Since we currently do not have a unified theory of gravity and quantum mechanics, we do not know which of these possibilities corresponds to reality. For the remainder of this book, we will ignore these questions and treat space-time as if it is differentiable and free of the complexities of quantum mechanics and quantum fields.

1.3 Tangents to Curves in One Dimension

Suppose we take the curve $y = 3x^2$, which has a gradient of 12 at the point $x = 2$, $y = 12$. The equation of a straight line with gradient $m = 12$ at point $(x_1, y_1) = (2, 12)$ is obtained by substituting it into $y_T - y_1 = m(x - x_1)$ and is $y - 12 = 12(x - 2)$ or $y_T = 12(x - 1)$. Both the curve and the straight line are plotted in Figure 1.1.

We can see that if we define $y' = y - 12x$, then the tangent in y' is horizontal at $x = 2$. This is because substituting into the tangent equation, we get $y' + 12x =$

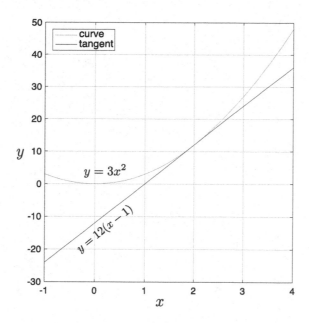

Figure 1.1 The curve $y = 3x^2$ and its tangent at $x = 2$, $y = 12(x - 1)$.

$12x - 1$ or just $y' = -12$. We perform the same linear transformation on the curve, so that $y' + 12x = 3x^2$, and we obtain $y' = 3x^2 - 12x$. Plotting $y'(x)$ for both curve and tangent leads to Figure 1.2.

We find that in the transformed variable y' the tangent has zero gradient, and the curve has a minimum at the point of intersection and rises quadratic in $x - 2$ on either side. In fact, we have $y' = 3(x - 2)^2$. In the coordinates x', y', related to x, y by the linear transformation $y' = y - 12x$, $x' = x - 2$, so that $y' = 3(x')^2$ in the transformed, primed coordinates, the deviation of the curve from a flat straight line in the neighbourhood of $x' = 0$ is quadratic in x'. This deviation vanishes faster than the deviation of x' from $x' = 0$. We could have chosen any other point on any curve, found the gradient of the curve at that point, and deduced a linear transformation that maps the curve onto one with zero gradient at that point and quadratic divergence from flatness moving away from that point, so long as the curve can be differentiated. This is guaranteed by the Taylor expansion. For any differentiable curve $y(x)$, we can write

$$y(x) = y(x_0) + \frac{dy}{dx}\bigg|_{x_0} (x - x_0) + \frac{1}{2} \frac{d^2y}{dx^2}\bigg|_{x_0} (x - x_0)^2 + \cdots. \tag{1.3}$$

The tangent curve at the point $x = x_0$ is

$$y_T(x) = y(x_0) + \frac{dy}{dx}\bigg|_{x_0} (x - x_0). \tag{1.4}$$

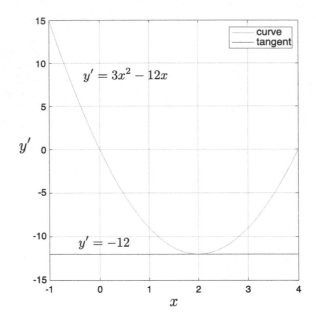

Figure 1.2 The same curve as in Figure 1.1 and its tangent at $x = 2$ following the linear transformation $y' = y - 12x$.

The transformation

$$y' = y - \frac{dy}{dx}\bigg|_{x_0} x \tag{1.5}$$

results in a curve $y'(x)$ having a local minimum or maximum at $x = x_0$. The difference between the curve $y'(x)$ and the tangent straight line of zero gradient are quadratic in the distance $x - x_0$ moving away from the point x_0. A linear transformation can be made on any differentiable function $y(x)$ to a coordinate y' that has a local minimum or maximum at any point x and is quadratic in departures from that point. This can be shown by substituting the transformation of Equation (1.5) into the expression for the Taylor expansion of the function $y(x)$ about the arbitrary point x_0:

$$y'(x) + \frac{dy}{dx}\bigg|_{x_0} x = y(x_0) + \frac{dy}{dx}\bigg|_{x_0} (x - x_0) + \frac{1}{2}\frac{d^2y}{dx^2}\bigg|_{x_0} (x - x_0)^2 + \cdots$$

$$y'(x) = y(x_0) - \frac{dy}{dx}\bigg|_{x_0} x_0 + \frac{1}{2}\frac{d^2y}{dx^2}\bigg|_{x_0} (x - x_0)^2 + \cdots. \tag{1.6}$$

The first two terms on the right are constants, and the third term is quadratic in $x - x_0$, so this curve has a local extremum at $x = x_0$ and deviates from the horizontal as a quadratic on either side of that point.

What have we learned from this? We have learned that for any differentiable curve, we can find a coordinate transformation into a coordinate system where that curve is flat at one point and quadratic in its deviations from flat as you move away from that point.

1.4 Curved Surfaces and Tangent Planes

We can also make tangents to two-dimensional curved planes. This is illustrated in Figure 1.3.

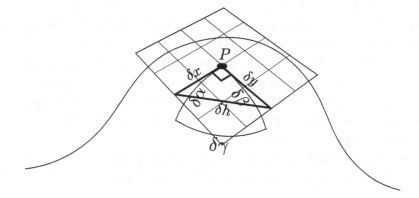

Figure 1.3 A curved surface with a tangent plane at point P. In the tangent plane, a small right-angled triangle has shorter sides δx and δy. In the curved space touching the plane, a curved triangle with edges corresponding to geodesics, or the straightest lines available in a curved space, has shorter sides $\delta \alpha$ and $\delta \beta$.

The flat plane touches the curved surface at point P. This time, instead of dealing with quantities outside the two-dimensional surfaces, let us confine ourselves to quantities that are within the surfaces. So, we are not going to determine the height z of the curved hill above some imagined flat base. Instead, we are going to start to think of ourselves as embedded in the two-dimensional curved surface, like a sort of two-dimensional ant, or alternatively embedded in the flat tangent plane. This is in preparation for considering the three-dimensional space and four-dimensional space-time, where we will not be able to imagine those spaces being embedded in some higher-dimensional space in the way we can imagine a two-dimensional surface embedded in a three-dimensional space.

One thing we can certainly do in the flat plane is to draw a right-angled triangle with its corner at the point P where it touches the curved sheet. We measure the two shorter sides of this triangle, δx and δy, and then the hypotenuse δh. We find that Pythagoras' theorem is obeyed, so that

$$\delta h^2 = \delta x^2 + \delta y^2. \tag{1.7}$$

We now ask the ants in the curved surface to try and reproduce this experiment in their curved space. Of course, they cannot make lines as straight as the ants in the tangent space. However, we can imagine that they can make their lines as straight as possible. For example, suppose they were to stretch an elastic band between two points in the curved surface. It would find the path for which its length was a minimum, and this would be as straight of a line as you can get confined to the curved surface. We will be more precise later in defining geodesics, but for now, it is enough to know that it is possible to determine the straightest possible path, even in a curved space. So, the ants do this for two lines that are at right angles at point P, and then they measure the distance between the two far ends of these lines, again along the geodesic joining the far ends, and they discover that the three lines are of lengths $\delta\alpha$, $\delta\beta$, and $\delta\gamma$. Do you think that these three lengths obey Pythagoras' theorem? It turns out that they do not. For example, suppose the curved surface was the surface of a sphere. Do the sides of right-angled spherical triangles obey Pythagoras' theorem? No they do not, and those of you who have studied spherical trigonometry in astronomy know that there are special rules governing the geometry of spherical triangles.

However, there is in fact a modified version of Pythagoras' theorem that triangles embedded in curved surfaces do obey. Here is a way of writing it:

$$\delta\gamma^2 = g_{11}\delta\alpha^2 + 2g_{12}\,\delta\alpha\,\delta\beta + g_{22}\delta\beta^2, \tag{1.8}$$

where g_{11}, g_{12}, and g_{22} are numerical factors known as metric coefficients, and they reflect the curvature of the surface at point P. Another way of writing this is

$$\delta\gamma^2 = \sum_{i=1}^{2}\sum_{j=1}^{2} g_{ij}\delta\xi^i\delta\xi^j, \tag{1.9}$$

where $\xi^1 = \alpha$, $\xi^2 = \beta$, and $g_{12} = g_{21}$. You are going to have to get used to indices written 'upstairs' in the position where sometimes you will see powers. Whether you are looking at a power or an index should be clear from the context.

Now that we have seen the modification of Pythagoras' theorem for a curved surface, we can write the ordinary Pythagoras' theorem in a flat surface in a way that makes it clear how they are related:

$$\delta h^2 = 1 \times \delta x^2 + 0 \times (\delta x)\,(\delta y) + 1 \times \delta y^2. \tag{1.10}$$

We can see that in the case where our curved surface is in fact flat as well, g_{11} and g_{22} would be 1, and g_{12} would be zero. Evidently, the quantities g_{11}, g_{12}, and g_{22} contain information about the curvature of the surface at point P. Equation (1.10) can also be written

$$\delta h^2 = \sum_{i=1}^{2} \sum_{j=1}^{2} \eta_{ij} \delta x^i \delta x^j, \tag{1.11}$$

where $x^1 = x$, $x^2 = y$, $\eta_{11} = \eta_{22} = 1$, and $\eta_{12} = \eta_{21} = 0$. The coefficients η_{ij} are particular cases of the coefficients g_{ij} that apply specifically to flat spaces described in Cartesian coordinates.

You can see how if the triangles were very small, the coefficients g_{ij} of the modified Pythagoras' theorem would be very close to the coefficients η_{ij} for a flat plane. In fact, exactly at point P, were our triangles to tend to zero size, there would be no difference between the geometry of the curved space and the geometry of the flat plane. This is calculus at work again – curves look straight at high magnification; curved surfaces look flat at high magnification too. However, as you move away from the point P where the two surfaces touch, and your triangles start to get bigger, you start to see discrepancies between the flat surface and the curved one. Those discrepancies show up as changes in the metric coefficients g_{ij}. We can write all this in terms of a Taylor expansion, this time a two-dimensional one, of the metric coefficients about the point (x^1, x^2) where the plane and the tangent surface intersect:

$$g_{ij}\left(x^1 + \delta x^1, x^2 + \delta x^2\right) = g_{ij}\left(x^1, x^2\right) + \frac{\partial g_{ij}}{\partial x^1} \delta x^1 + \frac{\partial g_{ij}}{\partial x^2} \delta x^2$$

$$+ \frac{1}{2}\frac{\partial^2 g_{ij}}{\partial (x^1)^2}\left(\delta x^1\right)^2 + \frac{1}{2}\frac{\partial^2 g_{ij}}{\partial (x^2)^2}\left(\delta x^2\right)^2$$

$$+ \frac{\partial^2 g_{ij}}{\partial x^1 \partial x^2}\left(\delta x^1\right)\left(\delta x^2\right) + \cdots$$

$$= \eta_{ij} + \frac{\partial g_{ij}}{\partial x^1} \delta x^1 + \frac{\partial g_{ij}}{\partial x^2} \delta x^2$$

$$+ \frac{1}{2}\frac{\partial^2 g_{ij}}{\partial (x^1)^2}\left(\delta x^1\right)^2 + \frac{1}{2}\frac{\partial^2 g_{ij}}{\partial (x^2)^2}\left(\delta x^2\right)^2$$

$$+ \frac{\partial^2 g_{ij}}{\partial x^1 \partial x^2}\left(\delta x^1\right)\left(\delta x^2\right) + \cdots. \tag{1.12}$$

Importantly, all the derivatives of the metric coefficients g_{ij} appearing in Equations (1.12) are evaluated at the point of contact (x^1, x^2) between the curved space and the tangent plane. In the second equality, I have substituted $g_{ij}(x^1, x^2) = \eta_{ij}$.

Recall that in one dimension, I can make a coordinate transformation such that any given point on a curve has zero gradient in the transformed coordinates. In the same way, in two dimensions, I can always find a coordinate system where the two first derivatives of g_{ij} with respect to the two new coordinates are zero,

$$\frac{\partial g_{ij}}{\partial y^1}(y^1, y^2) = \frac{\partial g_{ij}}{\partial y^2}(y^1, y^2) = 0. \tag{1.13}$$

The coordinates for the contact point are (y^1, y^2) in the new coordinate system, and we can therefore write

$$g_{ij}(y^1 + \delta y^1, y^2 + \delta y^2) = \eta_{ij} + \frac{1}{2}\frac{\partial^2 g_{ij}}{\partial (y^1)^2}(\delta y^1)^2 + \frac{1}{2}\frac{\partial^2 g_{ij}}{\partial (y^2)^2}(\delta y^2)^2$$

$$+ \frac{\partial^2 g_{ij}}{\partial y^1 \partial y^2}(\delta y^1)(\delta y^2) + \cdots. \tag{1.14}$$

This equation can also be written using two summation signs:

$$g_{ij}(y^1 + \delta y^1, y^2 + \delta y^2) = \eta_{ij} + \frac{1}{2}\sum_{j=1}^{2}\sum_{k=1}^{2}\frac{\partial^2 g_{ij}}{\partial y^j \partial y^k}\delta y^j \delta y^l + \cdots. \tag{1.15}$$

Though there are four terms in the double sum, the two where $j \neq k$ are equal. Therefore, in two dimensions, there are three independent second derivatives of g_{ij} in the first non-zero corrections to the flat space metric η_{ij}. This equation is only true in the special coordinates where the first derivatives of the metric with respect to displacements from the contact point are zero. We refer to coordinate systems of this type as locally coincident coordinates, or, since this is a bit of a mouthful, somewhat humorously as 'pigeon' coordinates. The nickname pigeon will be justified in the next couple of sections, where we next consider how these ideas apply to four-dimensional space-time.

1.5 Four-Dimensional Space-Time

All of you have encountered four-dimensional space-time in special relativity, and before we introduce the ideas of Einstein, we need to figure out how to carry over the ideas of curved spaces in two dimensions to possibly curved space-times in four! Unfortunately, it is impossible to visualise four dimensions, but let us just think about what the four-dimensional space-time equivalents of the curved surfaces we have been discussing might be.

In special relativity, we neglected the action of forces on observers, and therefore bodies in special relativity tend to move at constant velocity. We thought of those non-accelerating observers as having their own reference frames, coordinate systems in which the position and time coordinates of events are recorded. There are other observers moving with respect to any given observer who have their own reference frames, and the coordinates of the same event as measured by different observers are related by the Lorentz transforms. For example, if two observers have a relative velocity between them of magnitude v directed along the common x axis, and if their coordinate systems are such that their y and z axes are parallel, then the coordinates of the same event in their two different frames of reference are related by

$$ct' = \gamma(ct - \beta x)$$
$$x' = \gamma(x - \beta ct)$$
$$y' = y$$
$$z' = z, \tag{1.16}$$

where $\gamma = 1/\sqrt{1 - \beta^2}$ and $\beta = v/c$. As neither of the two observers is accelerating, and we are not thinking that there is any possibility that space-time is curved in some way, it is fair to assume that the spatial components of these coordinate systems can extend over all space. As we know that we are about to introduce gravity, we might first stop talking about possibly large coordinates x and instead talk about the displacements in space-time between close-by events, so that

$$c\,dt' = \gamma(c\,dt - \beta\,dx)$$
$$dx' = \gamma(dx - \beta c\,dt)$$
$$dy' = dy$$
$$dz' = dz. \tag{1.17}$$

If you calculate minus the square of the first equation and add the sum of the squares of the other three, you are left with the relationship

$$-c^2\,dt'^2 + dx'^2 + dy'^2 + dz'^2 = -c^2\,dt^2 + dx^2 + dy^2 + dz^2. \tag{1.18}$$

This combination of squares with the minus sign on the time component seems to be independent of the velocity of the coordinate system with respect to the events. The analogy in two- or three-dimensional space is the length of a vector, which does not change when you transform coordinate systems by rotating the axes. This combination of derivatives is called the Lorentz invariant interval,

$$ds^2 = -c^2\,dt^2 + dx^2 + dy^2 + dz^2. \tag{1.19}$$

Again, using the summation convention, we can express this as

$$ds^2 = \sum_{\alpha=0}^{3}\sum_{\beta=0}^{3} \eta_{\alpha\beta}\,dx^\alpha\,dx^\beta. \tag{1.20}$$

Here we have extended the definition of the η_{ij} of our previous two-dimensional space to $\eta_{\alpha\beta}$ in four-dimensional space-time. The components with $\alpha \neq \beta$ are all zero, The zero-zero component is $\eta_{00} = -1$, and $\eta_{11} = \eta_{22} = \eta_{33} = +1$. I should note here that this choice of the numerical values of these components is a specific convention. Though I will stick to this convention throughout this book, other conventions are in use in other texts and sources. The interested reader is referred to Section 9.3.1.

We can carry over the geometric ideas of Section 1.4 to four space-time dimensions. There are going to be some observers; let us carry on calling them inertial observers, for whom the metric coefficients are $\eta_{\alpha\beta}$. We will discuss which observers are these in the next section; for now, just take it as an assumption. Now let us say some other observers manouver themselves alongside this first class of observers, so that for a single time instant, the origins of the coordinate systems of the two sets of observers coincide. This space-time point is the point of contact between the flat space-time of the inertial observers and the curved space-time of the other, locally coincident observers, or using the silly nickname introduced in Section 1.4, pigeon observers. The connection between the metric coefficients for inertial and locally coincident, or 'pigeon' observers are given by the extension of Equation (1.14) to four space-time dimensions:

$$
g_{\alpha\beta}\left(x^0 + \delta x^0, x^1 + \delta x^1, x^2 + \delta x^2, x^3 + \delta x^3\right) = \eta_{\alpha\beta}
$$

$$
+ \frac{1}{2}\frac{\partial^2 g_{\alpha\beta}}{\partial (x^0)^2}\delta\left(x^0\right)^2 + \frac{1}{2}\frac{\partial^2 g_{\alpha\beta}}{\partial (x^1)^2}\delta\left(x^1\right)^2 + \frac{1}{2}\frac{\partial^2 g_{\alpha\beta}}{\partial (x^2)^2}\delta\left(x^2\right)^2 + \frac{1}{2}\frac{\partial^2 g_{\alpha\beta}}{\partial (x^3)^2}\delta\left(x^3\right)^2
$$

$$
+ \frac{\partial^2 g_{\alpha\beta}}{\partial x^0 \partial x^1}\delta x^0 \delta x^1 + \frac{\partial^2 g_{\alpha\beta}}{\partial x^1 \partial x^2}\delta x^1 \delta x^2 + \frac{\partial^2 g_{\alpha\beta}}{\partial x^2 \partial x^3}\delta x^2 \delta x^3
$$

$$
+ \frac{\partial^2 g_{\alpha\beta}}{\partial x^0 \partial x^2}\delta x^0 \delta x^2 + \frac{\partial^2 g_{\alpha\beta}}{\partial x^1 \partial x^3}\delta x^1 \delta x^3 + \frac{\partial^2 g_{\alpha\beta}}{\partial x^0 \partial x^3}\delta x^0 \delta x^3. \tag{1.21}
$$

Notice that there are ten second derivatives of the $g_{\alpha\beta}$ with respect to displacements in Equation (1.21), whereas there were three independent second derivatives in Equation (1.14) for the two-dimensional space. This number, 10, of independent coefficients, should be remembered by the student for later on, in Chapter 3, when we start to look at how energy, momentum, and stress in space-time are encoded in mathematics. It will turn out that this number leads us to an inspired guess about the connection between energy, momentum, and mass and curved spaces and that this guess leads us later on to Einstein's gravitational equations, the foundations of general relativity.

In summation convention, Equation (1.21) can be written as

$$
g_{\alpha\beta}\left(x^0 + \delta x^0, x^1 + \delta x^1, x^2 + \delta x^2, x^3 + \delta x^3\right)
$$

$$
= \eta_{\alpha\beta} + \frac{1}{2}\sum_{\mu=0}^{3}\sum_{\nu=0}^{3}\frac{\partial^2 g_{\alpha\beta}}{\partial x^\mu \partial x^\nu}\delta x^\mu \delta x^\nu + \cdots . \tag{1.22}
$$

This is of exactly the same form as Equation (1.15). You can begin to see that use of index notation allows us to make connections between the same concepts in different numbers of dimensions. This is one of the powerful features of index notation.

1.6 Einstein's Principle of Equivalence

The existence of inertial observers is essential to making progress in solving relativity problems. The reason is that we understand how to formulate problems in a flat space and flat space-time. We understand how to do things like take dot products of vectors or the divergences of fields. We have built up a lot of mathematical capability based on our intuition about how things behave in spaces like this. But it is unavoidable that in relativity, some observers are going to be non-inertial, and for those observers, the coordinate systems we need to describe their motions are going to be peculiar. The question is which observers are non-inertial and which observers are inertial?

The answer is contained a genuinely wonderful piece of insight by Einstein, which is called the principle of equivalence. Here is a statement of Einstein's principle of equivalence:

An observer freely falling in a gravitational field making local measurements of an object also in free fall obtains results the same as would be obtained were the observer and the object moving at constant velocity in a gravitational field free region.

Einstein's principle of equivalence says that if you are falling freely in a gravitational field, then things appear, locally, as they would if you were moving at a constant velocity in a region with no gravitational field. This is actually a complete surprise, or at least it was to me when I first realised what it meant. In a gravitational field, freely falling objects are in inertial frames! To them, the geometry of the Universe is that of a flat space-time, and to them, locally, there is no gravity.

This initially counter-intuitive statement is actually perfectly sensible. Imagine that you are standing in an elevator, and some malicious person cuts the cable, so that you start to fall. Inside the elevator, you would seem weightless, because both you and the elevator body are falling with the same gravitational acceleration. After all, you both started with the same initial velocity at the time the cable was cut, and you and the elevator are both falling in the same gravitational field. If there were somebody inside the elevator with you, and you started a game of catch, then as you threw a ball back and forth between you, it would travel in a straight line, at a constant velocity, in the coordinate system you share with the elevator. If you were to hold out a mass attached to the end of a spring, then the spring would not be stretched because there would be no force on the mass; it too is weightless. Of course, all the aforementioned objects are in free fall in a gravitational field, but that does not matter in the frame of reference defined by the walls of the elevator. Inside that box, there is no physical manifestation of the gravitational field, and everything appears inertial.

As a consequence, we can write down an expression for the interval ds^2 between two events separated by coordinate displacement given by Equation (1.19):

$$ds^2 = -c^2\,dt^2 + dx^2 + dy^2 + dz^2. \tag{1.23}$$

There is no local hint of an effect of gravity in this equation, because to these freely falling observers, there is no local manifestation of any gravitational field!

What about observers who are not freely falling in a gravitational field? These are the observers for whom space-time is curved. A sub-class of these observers are the locally coincident, or pigeon observers; see metric coefficients of the form given in Equations (1.21) and (1.22), where the first derivatives of $g_{\alpha\beta}$ with respect to displacements are zero. The nickname 'pigeon' for these observers comes from this analogy. Suppose I am in free fall. I am therefore an inertial observer by the principle of equivalence. Along comes a pigeon, who uses its wings to manouver right up to me so that for an instant, the origin of its coordinates is spatially coincident with mine and moving at the same velocity. It is this pigeon whose coordinate system is locally coincident with mine, and it is an observer for whom the first derivatives of $g_{\alpha\beta}$ with respect to displacements away from the point of coincidence are zero. Hence the name pigeon coordinates. We will come back to this image, terminology, and these coordinate systems later in the book, particularly in Chapter 5.

So, to summarise, observers not in free fall witness the curved trajectories of objects in free fall under gravity. So, for these observers, we will require the more complicated coordinate systems that will contain within them the manifestation of gravitational fields. We will study, for example, gravity in the vicinity of a spherically symmetric mass distribution due to a non-spinning black hole and approximately present also in planets and stars that have approximate spherical symmetry. We will discover that the coefficients of the metric in these cases contain factors of Newton's gravitational constant, so that if you are standing on a planet and therefore not freely falling in a gravitational field, then you can feel the effects of the gravitational force.

There is an analogy here with pseudoforces in accelerating frames of reference. When you are standing on a roundabout in a playground, you feel drawn towards the edge of the roundabout. This feels like a real force as you are rotating, but actually it is a pseudoforce, a manifestation of the fact that you are accelerating. But, if you were to hold out a mass on a spring and let the mass go, you would feel the mass pulling radially outwards, and the centrifugal pseudoforce would be real to you in your accelerating coordinate system. In the same way, by Einstein's principle of equivalence, when you are not in free fall, you are not in an inertial frame of reference. All non-inertial frames of reference are considered to be accelerating frames, and in accelerating frames, you get pseudoforces. Yes, in a way, gravity is a pseudoforce. Another way of thinking about this: The only way to resist free fall in a gravitational field is to be fixed against free fall by the action of some other non-gravitational force. So, in a sense, gravity is only experienced by virtue of the other forces stopping us from being in free fall. It is a most interesting piece of physics, and I hope you will enjoy learning about it.

1.7 The Einstein Summation Convention

You will have noticed several expressions where coordinate indices are summed over the number of dimensions in the space where they are defined – so that i was summed from 1 to 2 in Section 1.4 and α was summed from 0 to 3 in Section 1.5. It gets tiresome to keep writing all those summation signs, especially as they often occur in pairs, so in the remaining chapters of this book, we follow the Einstein summation convention, whereby if an index is repeated twice on the same side of an equals sign, then the terms containing the index are summed, with each term in the sum having the repeated index taking all the values it can take. Usually, I will use Latin indices i, j, k, etc. for indices that sum over up to three dimensions in space and Greek indices α, β, etc. for indices that sum over time (time is always the 0th coordinate) and the three spatial dimensions. Under this convention, here are some of the equations that have appeared in this chapter:

$$ds^2 = \eta_{\alpha\beta}\, dx^\alpha\, dx^\beta \tag{1.24}$$

$$g_{\alpha\beta}\left(x^0 + \delta x^0,\, x^1 + \delta x^1,\, x^2 + \delta x^2,\, x^3 + \delta x^3\right)$$
$$= \eta_{\alpha\beta} + \frac{1}{2}\frac{\partial^2 g_{\alpha\beta}}{\partial x^\mu \partial x^\nu}\delta x^\mu \delta x^\nu. \tag{1.25}$$

The latter is the same equation as (1.21) but without all the second-order correction terms written out explicitly.

Another commonly used convention is that when something like the components $g_{\alpha\beta}$ in Equation (1.25) depends on a four-dimensional position vector, that vector is just expressed with the letter symbol only. So Equation (1.25) becomes much simpler,

$$g_{\alpha\beta}(x + \delta x) = \eta_{\alpha\beta} + \frac{1}{2}\frac{\partial^2 g_{\alpha\beta}}{\partial x^\mu \partial x^\nu}\delta x^\mu \delta x^\nu. \tag{1.26}$$

Notice that the indices α and β are not summed over, because they only occur on one side of the equation, but the indices μ and ν are summed over because each occurs twice on the right-hand side of the equals sign.

1 The Principle of Equivalence

1.8 Diagnostic Quiz

These questions are intended as practice and revision for students taking courses based on this book in mathematical techniques that will be used in the course. Answer as many as you can.

For the following, consider r and θ to be functions of t. For parts (k) to (o) you can take dr/dt and $d\theta/dt$ to be independent of r and θ in the sense that the partial derivatives of dr/dt and $d\theta/dt$ with respect to r and θ are zero. This mirrors the practical situation that frequently arises in general relativity problems. Evaluate:

a) $\dfrac{d}{dt}\left(e^{\theta}\right)$

b) $\dfrac{d}{dt}\left(e^{-r^2}\right)$

c) $\dfrac{d}{dt}\left(\cos(r\theta)\right)$

d) $\dfrac{d}{dt}\left(\cosh\left(\dfrac{r}{\theta}\right)\right)$

e) $\dfrac{d}{dt}\left(r\sinh(e^{\theta^2})\right)$

f) $\dfrac{d}{dt}\left(r\dfrac{d\theta}{dt}\right)$

g) $\dfrac{d}{dt}\left(r^2\dfrac{dr}{dt}\dfrac{d\theta}{dt}\right)$

h) $\dfrac{d}{dt}\left(r^2\left(\dfrac{d\theta}{dt}\right)^3\right)$

i) $\dfrac{d}{dt}\left(\dfrac{d\theta/dt}{dr/dt}\right)$

j) $\dfrac{d}{dt}\left(e^{(d\theta/dt)/(dr/dt)}\right)$

k) $\dfrac{\partial}{\partial r}(r\theta)$

l) $\dfrac{d}{dt}\dfrac{\partial}{\partial r}\left(\dfrac{r}{\theta}\right)$

m) $\dfrac{\partial}{\partial r}\dfrac{d}{dt}\left(\dfrac{r}{\theta}\right)$

n) $\dfrac{d}{dt}\dfrac{\partial}{\partial r}\left(\dfrac{r}{\theta^3}\right)$

o) $\dfrac{d}{dt}\dfrac{\partial}{\partial r}\dfrac{\partial}{\partial \theta}(\theta^3 r^{-3})$.

Find the inverse of the following matrices ($\gamma = 1/\sqrt{1-\beta^2}$):

p) $\begin{pmatrix} \cos\theta & \sin\theta \\ -\sin\theta & \cos\theta \end{pmatrix}$

q) $\begin{pmatrix} A & 0 & 0 \\ 0 & B & 0 \\ 0 & 0 & C \end{pmatrix}$

r) $\begin{pmatrix} A & 0 & 0 \\ 0 & P & Q \\ 0 & R & S \end{pmatrix}$

s) $\begin{pmatrix} \gamma & 0 & 0 & -\beta\gamma \\ 0 & 1 & 0 & 0 \\ 0 & 0 & 1 & 0 \\ -\beta\gamma & 0 & 0 & \gamma \end{pmatrix}$.

Solve, or write down the solutions to, the following differential equations, for general real solutions (A and B are constants, y is a real function of x):

t) $\dfrac{dy}{dx} = -A^2$

u) $\dfrac{dy}{dx} = -A^2 x$

v) $\dfrac{dy}{dx} = -A^2 y$

w) $\dfrac{d^2 y}{dx^2} = -A^2 y$

x) $\dfrac{d^2 y}{dx^2} = A^2 y$

y) $\dfrac{dy}{dx} = \dfrac{1}{B^2 + x^2}$

z) $\dfrac{dy}{dx} = \dfrac{B}{B+x}$ aa) $\dfrac{dy}{dx} = \dfrac{x}{B+x}$ ab) $\dfrac{dy}{dx} = \dfrac{1}{\sqrt{B^2 - x^2}}$

ac) $\dfrac{d^2 y}{dx^2} = y + 1.$

Sketch the following functions of x, where a, b, and c are positive real constants:

ad) $y(x) = \ln(x)$ ae) $y(x) = \ln(x-1)$ af) $y(x) = ae^{-bx}\cos(cx)$

ag) $y(x) = \dfrac{x+1}{x-2}.$

1.9 Problems

1.1   Starting from the definition of the derivative in Equation (1.1), show that the first derivative of x^n, where n is a positive integer, is nx^{n-1}.

1.2   A triangle drawn on the surface of a sphere of radius R consists of three segments of great circles that join at three vertices. It is a right-angled triangle if one of the three interior angles of the figure is $90°$. It is a small spherical triangle if the lengths of the three segments are all much less than the radius R. If the three sides are of length a, b, and c, and if there is a right angle between the sides of lengths b and c, then the cosine rule of spherical trigonometry simplifies to $\cos a = (\cos b) \times (\cos c)$.

(a) Show that to second order in a, b, and c or products of these, we have $a^2 \simeq b^2 + c^2$.

(b) Show that to fourth order the explicit form of the modified Pythagoras' theorem is

$$a^2 \simeq b^2 + c^2 - \frac{1}{3}b^2 c^2.$$

Hint: when you encounter a term in a^4, you will need to re-express this in terms of terms in b and c. You can substitute the second order result from part (a) so long as you can argue that any other terms beyond this approximation result in corrections at higher order than b^4, c^4, or $b^2 c^2$ and can therefore be neglected at fourth order.

1.3   Physics problems on spheres are commonly done using spherical polar coordinates at fixed radius, (θ, ϕ), where θ is the colatitude, the angle between the line joining the sphere centre and the north pole and the line joining the sphere centre and the point in question, and ϕ is the azimuthal angle between an arbitrary great circle running through the north and south poles (in the case of the Earth, this arbitrary great circle also passes through Greenwich) on the sphere and the plane running through the north and south poles and the point in question. A three-sided figure on the surface of a sphere is constructed by starting at the north pole, then descending at constant ϕ to some value of θ, next moving at fixed θ through an interval of ϕ, so remaining at constant latitude and moving east or west, then finally travelling back at fixed ϕ to the north pole. Is this three-sided figure a spherical triangle? If not, why not?

1.4 Generalising Equations (1.14) in two dimensions and (1.21) in four dimensions, how many independent second-order derivatives are there of the metric coefficients in D dimensions?

1.5 A lift is falling freely under gravity. An observer in the lift has set up a laser pointer attached to the wall of the lift pointing in her coordinate systems at right angles so that the emerging beam is emitted normal to the wall. Bearing in mind the principle of equivalence, what is the pathway of the photons across the lift? What would an observer outside the lift standing on the Earth's surface observe about the light beam as the lift falls? It is a glass lift, so both observers can see the beam.

1.6 Specialist aircraft such as the KC-105 provided early trainee astronauts with a zero gravity environment. Incidentally, the KC-105 was nicknamed 'the vomit comet'! Apparently, the first experience of a zero gravity environment can be uncomfortable. A KC-105 starts its zero gravity flight segment at a height h flying at an initial velocity $\vec{v}$ directed 45° above the horizontal. The passengers remain in zero gravity until the plane again reaches the same altitude h. Take the x and y axes to be parallel and perpendicular to the Earth's surface, respectively.

(a) Sketch the trajectory of the aircraft during its zero gravity segment.

(b) Assuming that the maximum vertical displacement of the plane is much less than the radius of the Earth, give formulae for the horizontal and vertical components of the aircraft displacement during the zero gravity flight segment. Take the start of the zero gravity segment to be at time $t = 0$ and position $x = 0$, $y = h$, the subsequent motion to be in the XY plane, and the horizontal component of the initial velocity to be in the direction of increasing x. Express your answers in terms of the quantities defined above and the local gravitational acceleration magnitude g directed downwards.

2

Tensors

2.1 Curvilinear Coordinates

In Chapter 1, a picture emerged of two distinct classes of observers, inertial and accelerating. In special relativity, the inertial observers are moving at a constant velocity in the absence of any external forces. In general relativity, the inertial observers are in free fall in a gravitational field. These two definitions of inertial are consistent because if a freely falling observer moves into a region of zero gravitational field, then he will move at a constant velocity relative to special-relativistic inertial observers.

In the presence of a gravitational field, for an observer not in free fall, gravity manifests itself as curvature of space-time in his frame of reference. The metric coefficients develop terms containing Newton's gravitational constant corresponding to a local gravitational field induced, for example, by massive objects in the vicinity.

This is a compelling picture, and indeed it is correct. However, there is a subtlety. If in an observer's coordinate system the metric coefficients are determined to be η_{ij} (in a space) or $\eta_{\alpha\beta}$ (in space-time), then this is sufficient to conclude that the observer is in an inertial frame. However, the opposite is not true. If in an observer's coordinate system the metric coefficients are not η_{ij} (in space) or $\eta_{\alpha\beta}$ (in space-time), then this is insufficient to determine that this observer is in a curved space (in space) or is in an accelerating frame of reference (in space-time).

The reason for this has to do with the set of possible coordinate systems. A coordinate system consists of covering a space with a grid of lines. Along each of these lines, all but one of the coordinates are constant. So, in the simplest example, Cartesian coordinates in two spatial dimensions, the vertical grid lines are lines of constant x, and the horizontal grid lines are lines of constant y. A point in space is specified by a value (x, y) of these coordinates.

In the case of rectilinear coordinate systems, which we have been considering so far, and of which Cartesian coordinates are an example, the grid lines run along geodesics – straight lines in flat space, the lines corresponding to stretched elastic bands in a curved space, or the pathways of freely falling observers in space-time. However, coordinate systems are used, very frequently, where the grid lines do not run along geodesics. Such coordinate systems are called curvilinear. We will see that in curvilinear coordinate systems, the metric coefficients differ from the η_{ij} (in space) or the $\eta_{\alpha\beta}$ (in space-time), even when the space is flat, or in space-time even when the observer's frame of reference is inertial.

The simplest example is the flat two-dimensional Euclidean space in plane polar coordinates, so let us see what the metric coefficients look like in that coordinate system.

2.2 Metric Coefficients in Plane Polar Coordinates

In plane polar coordinates, a point is specified by its distance from the origin r and the angle a straight line drawn from that point to the origin makes with the positive x-axis. The plane polar coordinates of a point (r, θ) are therefore related to the Cartesian coordinates of the same point (x, y) by

$$x = r \cos \theta$$
$$y = r \sin \theta. \tag{2.1}$$

In Cartesian coordinates, you may be used to writing vectors by specifying the components of their displacements, so, for example,

$$\vec{r} = \begin{pmatrix} 3 \\ -2 \end{pmatrix}. \tag{2.2}$$

The components are the numbers of units of the two Cartesian unit vectors you have to translate through to have the sum of these three vector displacements be equivalent to the displacement $\vec{r}$ represented by the vector. So, we can also write

$$\vec{r} = 3\hat{\mathbf{i}} - 2\hat{\mathbf{j}}. \tag{2.3}$$

We will want to refer to the basis vectors collectively using index notation, so let us write $\vec{e}_x = \hat{\mathbf{i}}$ and $\vec{e}_y = \hat{\mathbf{j}}$, and also later when we need the third Cartesian axis, $\vec{e}_z = \hat{\mathbf{k}}$. Collectively, we could refer to the basis vectors as $\vec{e}_i$. Also, if we write the components of our vector displacement as x^i, where $x^1 = x$ and $x^2 = y$, then we can use the Einstein summation convention to write a vector $\vec{r}$ in terms of its Cartesian components as

$$\vec{r} = x^i \vec{e}_i. \tag{2.4}$$

Now let us think about how things work in polar coordinates. It is certainly okay to specify the position of a point by giving its polar coordinates (r, θ). However, is it okay to describe a vector $\vec{r}$ as $\vec{r} = r\vec{e}_r + \theta\vec{e}_\theta$? Actually, it is not okay. This is because, unlike the Cartesian basis vectors $\vec{e}_x$ and $\vec{e}_y$, the basis vectors in plane polar coordinates, $\vec{e}_r$ and $\vec{e}_\theta$, pointing as they do in the directions of increasing r and θ, respectively, vary from point to point in the space! So, if you write $\vec{r} = r\vec{e}_r + \theta\vec{e}_\theta$ for a large displacement $\vec{r}$ between two points, then which basis vectors $\vec{e}_r$ and $\vec{e}_\theta$ do you mean? The ones at the start of the vector or the ones at the end?

The resolution to this problem is to move from thinking about large vector displacements $\vec{r}$ to small vector displacements $d\vec{r}$. This means a vector displacement between two neighbouring points – points that are close together. In the two coordinate systems, we can write

$$d\vec{r} = dx\,\vec{e}_x + dy\,\vec{e}_y$$
$$= dr\,\vec{e}_r + d\theta\,\vec{e}_\theta. \tag{2.5}$$

The $\vec{e}_r$ and $\vec{e}_\theta$ will be almost the same at the two ends of the vector, but we nonetheless specifically define $\vec{e}_r$ and $\vec{e}_\theta$ as the basis vectors at the start of the vector $d\vec{r}$.

We have been talking a lot about these polar basis vectors, without working out expressions for them. We use the total derivative to break down the $d\vec{r}$ into components in both coordinate systems:

$$d\vec{r} = \left(\frac{\partial \vec{r}}{\partial x}\right)_y dx + \left(\frac{\partial \vec{r}}{\partial y}\right)_x dy$$
$$= \left(\frac{\partial \vec{r}}{\partial r}\right)_\theta dr + \left(\frac{\partial \vec{r}}{\partial \theta}\right)_r d\theta. \tag{2.6}$$

Comparing Equations (2.5) and (2.6), we can see that

$$\vec{e}_x = \left(\frac{\partial \vec{r}}{\partial x}\right)_y$$
$$\vec{e}_y = \left(\frac{\partial \vec{r}}{\partial y}\right)_x$$
$$\vec{e}_r = \left(\frac{\partial \vec{r}}{\partial r}\right)_\theta$$
$$\vec{e}_\theta = \left(\frac{\partial \vec{r}}{\partial \theta}\right)_r. \tag{2.7}$$

In general, if we have a basis vector $\vec{e}_i$ in some coordinate system, and we have an expression for the position of a point $\vec{r}$ in the space in Cartesian coordinates, then we can write

$$\vec{e}_i = \left(\frac{\partial \vec{r}}{\partial x^i} \right)_{x^{j \neq i}}, \tag{2.8}$$

where x^i is any one of the coordinates in that system. Henceforth we will omit specification of the coordinates $x^{j \neq i}$ held fixed when partially differentiating with respect to x^i. The rule is that all the other coordinates of the system of which x^i is a member are held fixed when partially differentiating with respect to x^i. You can check that this works for Cartesian coordinates by starting with $\vec{r} = x\vec{e}_x + y\vec{e}_y$ and calculating the partial derivatives with respect to x and y to obtain $\vec{e}_x$ and $\vec{e}_y$.

Let us apply this to work out expressions for the basis vectors in polar coordinates. Once again, write $\vec{r} = x\vec{e}_x + y\vec{e}_y$ but this time substitute for x and y with $x = r \cos \theta$ and $y = r \sin \theta$. We obtain

$$\vec{r} = r \cos \theta \vec{e}_x + r \sin \theta \vec{e}_y. \tag{2.9}$$

We can then derive expressions for $\vec{e}_r$ and $\vec{e}_\theta$ by taking the partial derivatives with respect to r and θ. Note that $\vec{e}_x$ and $\vec{e}_y$ have the same magnitude and direction everywhere – a special property of Cartesian coordinates – so that they are constants with respect to taking derivatives:

$$\vec{e}_r = \left(\frac{\partial \vec{r}}{\partial r} \right)_\theta = \cos \theta \vec{e}_x + \sin \theta \vec{e}_y$$

$$\vec{e}_\theta = \left(\frac{\partial \vec{r}}{\partial \theta} \right)_r = -r \sin \theta \vec{e}_x + r \cos \theta \vec{e}_y. \tag{2.10}$$

Notice that $\vec{e}_\theta$ is not of unit length! The modern convention is that basis vectors do not have to be length one, as they are in Cartesian coordinates. This makes it simpler to do algebra but adds an extra step in calculations where you want to actually work out the components of a vector, say an electric field. These components are expressed in multiples of the basis vectors in the coordinate system being used, so you need to multiply the component by the length of the basis vector to get the correct answer for the magnitude of the physical quantity. Also note that many quite modern maths methods books persist in using basis vectors of unit length, resulting in the insertion of extra scale factors. I will stick with the definition of a basis vector given by $\vec{e}_i = \partial \vec{r} / \partial x^i$, even if it does mean that basis vectors in general have a position-dependent length and direction.

Now that we have defined the basis vectors in plane polar coordinates, we can work out the metric coefficients. Start from the fact that the length of the vector $d\vec{r}$ is going to be the same whichever coordinate system you write it in. So, let us write it in plane polar coordinates. The square of this length is

$$|d\vec{r}|^2 = d\vec{r} \cdot d\vec{r} = (dr\,\vec{e}_r + d\theta\,\vec{e}_\theta) \cdot (dr\,\vec{e}_r + d\theta\,\vec{e}_\theta)$$
$$= dr^2\,\vec{e}_r \cdot \vec{e}_r + 2\,dr\,d\theta\,\vec{e}_r \cdot \vec{e}_\theta + d\theta^2\,\vec{e}_\theta \cdot \vec{e}_\theta. \tag{2.11}$$

This expression leads us to a general way of defining the metric coefficients in a given coordinate system. They are just the dot products of the basis vectors amongst themselves. In the case of the plane polar coordinate system,

$$g_{rr} = \vec{e}_r \cdot \vec{e}_r \qquad\qquad g_{r\theta} = \vec{e}_r \cdot \vec{e}_\theta$$
$$g_{\theta r} = \vec{e}_\theta \cdot \vec{e}_r \qquad\qquad g_{\theta\theta} = \vec{e}_\theta \cdot \vec{e}_\theta. \tag{2.12}$$

To evaluate the dot products of $\vec{e}_r$ and $\vec{e}_\theta$, we use the expressions for them in Equations (2.10),

$$\vec{e}_r \cdot \vec{e}_r = (\cos\theta\,\vec{e}_x + \sin\theta\,\vec{e}_y) \cdot (\cos\theta\,\vec{e}_x + \sin\theta\,\vec{e}_y)$$
$$= \cos^2\theta + \sin^2\theta = 1, \tag{2.13}$$

where we have used $ex \cdot \vec{e}_x = \vec{e}_y \cdot \vec{e}_y = 1$ and $\vec{e}_x \cdot \vec{e}_y = \vec{e}_y \cdot \vec{e}_x = 0$. Similarly, for $\vec{e}_r \cdot \vec{e}_\theta$,

$$\vec{e}_r \cdot \vec{e}_\theta = (\cos\theta\,\vec{e}_x + \sin\theta\,\vec{e}_y) \cdot (-r\sin\theta\,\vec{e}_r + r\cos\theta\,\vec{e}_\theta)$$
$$= -r\sin\theta\cos\theta + r\sin\theta\cos\theta = 0, \tag{2.14}$$

which is not surprising since the directions of increasing r and increasing θ along which $\vec{e}_r$ and $\vec{e}_\theta$ lie are at right angles. Lastly, the interesting one

$$\vec{e}_\theta \cdot \vec{e}_\theta = (-r\sin\theta\,\vec{e}_r + r\cos\theta\,\vec{e}_\theta) \cdot (-r\sin\theta\,\vec{e}_r + r\cos\theta\,\vec{e}_\theta)$$
$$= r^2\sin^2\theta + r^2\cos^2\theta = r^2. \tag{2.15}$$

Substituting Equations (2.13), (2.14), and (2.15) into Equation (2.11) we arrive at

$$|d\vec{r}|^2 = dr^2 + r^2\,d\theta^2. \tag{2.16}$$

Recalling Equation (1.8), we see that in this coordinate system, we have, in matrix notation,

$$\begin{pmatrix} g_{rr} & g_{r\theta} \\ g_{\theta r} & g_{\theta\theta} \end{pmatrix} = \begin{pmatrix} 1 & 0 \\ 0 & r^2 \end{pmatrix}. \tag{2.17}$$

So in this simplest of spaces, the flat two-dimensional space, the choice of a curvilinear coordinate system results in metric coefficients that are not in the form η_{ij}. Put another way, we cannot tell by inspecting the metric coefficients that the space we are in is flat! It might be a curved space. It turns out that it is possible to work out whether the space you are in is flat or curved from the coefficients of the metric, but we will need a more subtle test than just inspection of the metric coefficients to do so.

2.3 A Closer Look at Polar Coordinates

It will turn out to be very useful to explore further the transformation between Cartesian and plane polar coordinate systems, because this will allow us to see more clearly how to handle inertial and accelerating observers in curved space-time. We go back to considering a small vector displacement $d\vec{r}$ between two neighbouring points in our flat space, corresponding to displacement components (dx, dy) in the Cartesian coordinate system and $(dr, d\theta)$ in the polar coordinate system. Using the total derivative, the two are related by

$$dr = \left(\frac{\partial r}{\partial x}\right)_y dx + \left(\frac{\partial r}{\partial y}\right)_x dy$$

$$d\theta = \left(\frac{\partial \theta}{\partial x}\right)_y dx + \left(\frac{\partial \theta}{\partial y}\right)_x dy. \tag{2.18}$$

In the Einstein summation convention, these transformations can be written more compactly as

$$dx^{j'} = \frac{\partial x^{j'}}{\partial x^i} dx^i, \tag{2.19}$$

where the polar coordinates are primed, so that $x^{1'} = r$ and $x^{2'} = \theta$, and the Cartesian coordinates are unprimed, so that $x^1 = x$ and $x^2 = y$. We will come back to this compact form at the end of the section. For now, let us work out the partial derivatives in the longhand version (2.18). Starting with $r = (x^2 + y^2)^{1/2}$, we have

$$\left(\frac{\partial r}{\partial x}\right)_y = \frac{x}{(x^2 + y^2)^{1/2}} = \frac{x}{r} = \cos\theta$$

$$\left(\frac{\partial r}{\partial y}\right)_x = \frac{y}{(x^2 + y^2)^{1/2}} = \frac{y}{r} = \sin\theta. \tag{2.20}$$

Next, starting with $\tan\theta = y/x$, we have

$$\sec^2\theta \left(\frac{\partial \theta}{\partial x}\right)_y = (1 + \tan^2\theta)\left(\frac{\partial \theta}{\partial x}\right)_y = \frac{-y}{x^2}$$

$$\left(1 + \frac{y^2}{x^2}\right)\left(\frac{\partial \theta}{\partial x}\right)_y = \left(\frac{x^2 + y^2}{x^2}\right)\left(\frac{\partial \theta}{\partial x}\right)_y = \frac{-y}{x^2}$$

$$\left(\frac{\partial \theta}{\partial x}\right)_y = \frac{-y}{r^2} = \frac{-\sin\theta}{r}, \tag{2.21}$$

and

$$\sec^2\theta \left(\frac{\partial \theta}{\partial y}\right)_x = \frac{1}{x}$$

$$\left(\frac{x^2+y^2}{x^2}\right)\left(\frac{\partial\theta}{\partial y}\right)_x = \frac{1}{x}$$

$$\left(\frac{\partial\theta}{\partial y}\right)_x = \frac{x}{r^2} = \frac{\cos\theta}{r}, \tag{2.22}$$

so Equations (2.18) become

$$dr = \cos\theta\,dx + \sin\theta\,dy$$

$$d\theta = \frac{-\sin\theta}{r}\,dx + \frac{\cos\theta}{r}\,dy. \tag{2.23}$$

We also have the equations for $\vec{e}_r$ and $\vec{e}_\theta$ in terms of $\vec{e}_x$ and $\vec{e}_y$ from Equations (2.10):

$$\vec{e}_r = \cos\theta\vec{e}_x + \sin\theta\vec{e}_y$$

$$\vec{e}_\theta = -r\sin\theta\vec{e}_x + r\cos\theta\vec{e}_y. \tag{2.24}$$

Let us write $d\vec{r}$ in terms of polar coordinates, substitute in from these equations, and see if we recover the right expression for Cartesian coordinates:

$$\begin{aligned}
d\vec{r} &= dr\,\vec{e}_r + d\theta\,\vec{e}_\theta \\
&= (\cos\theta\,dx + \sin\theta\,dy)(\cos\theta\vec{e}_x + \sin\theta\vec{e}_y) \\
&\quad + \left(\frac{-\sin\theta}{r}\,dx + \frac{\cos\theta}{r}\,dy\right)(-r\sin\theta\vec{e}_x + r\cos\theta\vec{e}_y) \\
&= \cos^2\theta\,dx\,\vec{e}_x + \cos\theta\sin\theta\,dx\,\vec{e}_y \\
&\quad + \sin\theta\cos\theta\,dy\,\vec{e}_x + \sin^2\theta\,dy\,\vec{e}_y \\
&\quad + \sin^2\theta\,dx\,\vec{e}_x - \sin\theta\cos\theta\,dx\,\vec{e}_y \\
&\quad - \cos\theta\sin\theta\,dy\,\vec{e}_x + \cos^2\theta\,dy\,\vec{e}_y \\
&= dx\,\vec{e}_x + dy\,\vec{e}_y. \tag{2.25}
\end{aligned}$$

This demonstrates that we have the correct transformation laws from the components dr and $d\theta$ of a displacement in polar coordinates into the components dx and dy of the same displacement in Cartesian coordinates, written previously in Equation (2.19), which is reproduced here:

$$dx^{j'} = \frac{\partial x^{j'}}{\partial x^i}\,dx^i, \tag{2.26}$$

where the primed coordinate system is the polar coordinates, and the unprimed coordinate system is Cartesian.

However, this is not the end of the story, because we also need to write down an index form for the transformations between the polar basis vectors $\vec{e}_r$ and $\vec{e}_\theta$ and the Cartesian basis vectors $\vec{e}_x$ and $\vec{e}_y$.

To work this out, we start again with our definition of a basis vector as the partial derivative of a general vector. In a primed coordinate system,

$$\vec{e}_{j'} = \frac{\partial \vec{r}}{\partial x^{j'}}. \tag{2.27}$$

We then use the chain rule for partial derivatives, the proof of which is set in Problem 2.4, to write

$$\vec{e}_{j'} = \frac{\partial \vec{r}}{\partial x^i}\frac{\partial x^i}{\partial x^{j'}}$$

$$= \frac{\partial x^i}{\partial x^{j'}}\vec{e}_i. \tag{2.28}$$

This is not the same transformation law as Equation (2.26). The partial derivatives are the other way up. We have therefore discovered that the basis vectors and the components of a small displacement transform using different rules but that those transformation rules bear a close resemblance to each other. In fact, they are both members of a larger family of objects, the so-called tensors, whose components transform according to one of a set of transformations. This will turn out to be a very powerful idea and an essential one for general relativity.

2.4 Tensors in Polar Coordinates

Let us review our study of polar coordinates. These coordinates are curvilinear, meaning that the grid of points that describes positions in space runs along lines that are not, specifically in the case of lines of constant r and varying θ, geodesics. In the case of ordinary flat space, geodesics mean straight lines. We can, at any point in space, work out the transformation to go from Cartesian coordinates to polar coordinates. To transform from a Cartesian description of a vector to a polar one, we need to transform both the basis vectors at that point and the components of the particular vector we want to transform. The vector itself remains unchanged. After all, it is just an arrow between two points. What changes is how you specify the vector.

The transformation equations for the vector components and for the basis vectors are given by Equations (2.26) and (2.28), reproduced here for convenience:

$$dx^{j'} = \frac{\partial x^{j'}}{\partial x^i}\,dx^i, \qquad\qquad \vec{e}_{j'} = \frac{\partial x^i}{\partial x^{j'}}\vec{e}_i. \tag{2.29}$$

We noticed that these transformation rules are different. This is why I was careful to write the components with their index upstairs and the basis vectors with their index downstairs. Objects that transform under local changes of coordinates like the components of vectors, the transformation law on the left, we will call tensors of rank $^1_0$, and we use the single upstairs index to remind us how they transform.

Objects that transform under local changes of coordinates like the basis vectors, the transformation law on the right, we will call tensors of rank $^0_1$, and we use the single downstairs index to remind us of how they transform.

We also figured out the metric coefficients in polar coordinates. Looking at Equation (2.12) and generalising using index notation, we obtain a general expression for the metric coefficients. We start by using the definition in the primed coordinate system,

$$g_{k'l'} = \vec{e}_{k'} \cdot \vec{e}_{l'}. \tag{2.30}$$

If we substitute the latter of Equations (2.29) into Equation (2.30), we arrive at the expression for the transformation of the metric coefficients to the primed coordinate system from the metric coefficients in the unprimed coordinate system:

$$
\begin{aligned}
g_{k'l'} &= \vec{e}_{k'} \cdot \vec{e}_{l'} \\
&= \frac{\partial x^i}{\partial x^{k'}} \frac{\partial x^j}{\partial x^{l'}} \vec{e}_i \cdot \vec{e}_j \\
&= \frac{\partial x^i}{\partial x^{k'}} \frac{\partial x^j}{\partial x^{l'}} g_{ij}.
\end{aligned} \tag{2.31}
$$

Objects that transform like the components of the metric we will call tensors of rank $^0_2$, and we use two downstairs indices to remind us of the transformation law – notice that the transformation components look like two copies of the same kind of partial derivative that was used to transform tensors of rank $^0_1$, one acting on each index in the transformed object.

2.5 Transformation Laws for the Components of Tensors of Arbitrary Rank

From Equations (2.29) and (2.31) you should be able to discern a pattern applicable to tensors having components with an arbitrary collection of upstairs and downstairs indices. For a tensor of rank ^m_n, the transformation law for the components is

$$A^{\alpha'_1 \alpha'_2 \cdots \alpha'_m}_{\beta'_1 \beta'_2 \cdots \beta'_n} = \frac{\partial x^{\alpha'_1}}{\partial x^{\mu_1}} \frac{\partial x^{\alpha'_2}}{\partial x^{\mu_2}} \cdots \frac{\partial x^{\alpha'_m}}{\partial x^{\mu_m}} \frac{\partial x^{\nu_1}}{\partial x^{\beta'_1}} \frac{\partial x^{\nu_2}}{\partial x^{\beta'_2}} \cdots \frac{\partial x^{\nu_n}}{\partial x^{\beta'_n}} A^{\mu_1 \mu_2 \cdots \mu_m}_{\nu_1 \nu_2 \cdots \nu_n}, \tag{2.32}$$

so that there are m partial derivatives with the primed components in the numerator and n partial derivatives with the primed components in the denominator. Such generality will not be needed often, but we include it here for completeness.

2.6 Tensor Analysis

Let us write down a small vector displacement again, in terms of components in some coordinate system, call it the unprimed coordinate system, and do some manipulation of it using the transformation rules that we summarised in Section 2.4:

$$d\vec{r} = dx^i \vec{e}_i. \tag{2.33}$$

Now according to our rules, we can write a transformation on $\vec{e}_i$ into a primed coordinate system (we studied the polar coordinate system, but perhaps it might be any other locally valid coordinate system as well):

$$\vec{e}_{j'} = \frac{\partial x^i}{\partial x^{j'}} \vec{e}_i. \tag{2.34}$$

There is nothing sacred about the primed and unprimed indices; we might just as well have applied the primed and unprimed labels the other way round. So, let us swap them and obtain the inverse transform from primed to unprimed coordinates:

$$\vec{e}_j = \frac{\partial x^{i'}}{\partial x^j} \vec{e}_{i'}. \tag{2.35}$$

We want to substitute this into Equation (2.33), but we have two problems: the i index is in the wrong place, but we cannot just rename j to i because we already have an i'. But i' is summed over, so we can give it a new symbol; as long as it stays primed, it still represents the same sum. You can always change the symbols used for indices that are repeated and imply a sum. Let us use k' instead:

$$\vec{e}_j = \frac{\partial x^{k'}}{\partial x^j} \vec{e}_{k'}. \tag{2.36}$$

Now that i' is not used, we can rename the j index as i to obtain exactly the same equation again, but it now has the i index in the same place as it is in Equation (2.33):

$$\vec{e}_i = \frac{\partial x^{k'}}{\partial x^i} \vec{e}_{k'}. \tag{2.37}$$

We now do the same thing for dx^i. Start with the expression for the transform of vector components from Equation (2.34):

$$dx^{j'} = \frac{\partial x^{j'}}{\partial x^i} dx^i. \tag{2.38}$$

Swap the primed indices and the unprimed ones:

$$dx^j = \frac{\partial x^j}{\partial x^{i'}} dx^{i'}. \tag{2.39}$$

Substitute a new index l for i and then use i instead of j:

$$dx^i = \frac{\partial x^i}{\partial x^{l'}} dx^{l'}. \tag{2.40}$$

Now substituting Equations (2.36) and (2.39) into Equation (2.33), we obtain

$$d\vec{r} = \frac{\partial x^i}{\partial x^{l'}} dx^{l'} \frac{\partial x^{k'}}{\partial x^i} \vec{e}_{k'}. \tag{2.41}$$

Now this contains the product of two partial derivatives, so we write them together. In index notation, you can always swap terms around as they just represent numerical components, so they all commute:

$$d\vec{r} = \frac{\partial x^i}{\partial x^{l'}} \frac{\partial x^{k'}}{\partial x^i} dx^{l'} \vec{e}_{k'}. \tag{2.42}$$

Now the two partial derivatives actually represent the sum as follows:

$$\frac{\partial x^i}{\partial x^{l'}} \frac{\partial x^{k'}}{\partial x^i} = \frac{\partial x^{k'}}{\partial x^1} \frac{\partial x^1}{\partial x^{l'}} + \frac{\partial x^{k'}}{\partial x^2} \frac{\partial x^2}{\partial x^{l'}}. \tag{2.43}$$

If z is a function of u and v, and both u and v are functions of x and y, then we can write the chain rules of partial derivatives, for example,

$$\frac{\partial z}{\partial x} = \frac{\partial z}{\partial u} \frac{\partial u}{\partial x} + \frac{\partial z}{\partial v} \frac{\partial v}{\partial x}, \tag{2.44}$$

and also

$$\frac{\partial z}{\partial y} = \frac{\partial z}{\partial u} \frac{\partial u}{\partial y} + \frac{\partial z}{\partial v} \frac{\partial v}{\partial y}, \tag{2.45}$$

so that the partial derivatives in Equation (2.41) simplify leading to

$$d\vec{r} = \frac{\partial x^{k'}}{\partial x^{l'}} dx^{l'} \vec{e}_{k'}. \tag{2.46}$$

The partial derivative is now of one coordinate in the primed system with respect to another. This is zero unless the two coordinates are the same one, so $\partial r/\partial r = 1$ but $\partial r/\partial \theta = 0$. Hence we can replace partial derivatives amongst components from the same coordinate system by Kronecker deltas. Equation (2.41) becomes

$$d\vec{r} = \delta^{k'}_{l'} dx^{l'} \vec{e}_{k'}, \tag{2.47}$$

where $\delta^{k'}_{l'}$ is zero unless $l' = k'$. So in the sums over all l' and k', only the terms with $l' = k'$ survive, so that we arrive at

$$d\vec{r} = dx^{k'} \vec{e}_{k'}. \tag{2.48}$$

Notice that there are now two expressions for the vector displacement $d\vec{r}$ in the two coordinate systems, and they look basically identical:

$$d\vec{r} = dx^i \vec{e}_i = dx^{k'} \vec{e}_{k'}. \tag{2.49}$$

Notice also that all the objects in the equation are tensors. You might wonder about $d\vec{r}$, Well, it is an invariant, since the vector arrow does not change just because somebody invents a new coordinate system to frame it in. It is called an invariant, and invariants are tensors of rank $^0_0$. They have no indices.

Here then is an example of the principle of tensor analysis

If you can write an equation between quantities measured locally in the neighbour-hood of some point entirely in terms of tensors, and you can show that the equation is true in one coordinate system, then it is true in all coordinate systems.

The principles of tensor analysis, and this rule, carry over lock stock and barrel into curved space-time in the presence of a gravitational field! Let us get an idea how that works.

2.7 Tensors in Space-Time

The great power of the tensor notation becomes apparent when we leave the restricted stage of a two-dimensional flat space and return to the space-time, which we actually inhabit. In special relativity, you encounter the Lorentz transformations between different inertial observers. Now that we are talking about gravitational fields, we are aware that the inertial observers may actually be in free fall in a gravitational field, but we consider the observers in the ensuing discussion to be located very close together, and only making local measurements, so that by the principle of equivalence the dynamics is the same as that for observers moving at constant velocities in the absence of a gravitational field. We consider two such observers with a relative velocity of $v = \beta c$, directed along their common x and x' axes. Both the observers measure two events close to each other in space-time, so that the coordinates of the displacement between them are $(c\,dt, dx, dy, dz)$ to one of the observers and $(c\,dt', dx', dy', dz')$ to the other, where the primed observer is moving in the positive x direction with respect to the unprimed one. The coordinates of the displacement are related by a Lorentz transformation, so that

$$c\,dt' = \gamma(c\,dt - \beta\,dx)$$
$$dx' = \gamma(dx - \beta\,c\,dt)$$
$$dy' = dy$$
$$dz' = dz. \tag{2.50}$$

If we think of Lorentz transformations as an example of a coordinate transformation between different coordinate systems in space-time, then we can carry over the same index notation we used to express transformations between Cartesian and plane polar coordinates. Let $dx^0 = c\,dt$, $dx^1 = dx$, $dx^2 = dy$, $dx^3 = dz$, and similarly for the primed coordinates. Then we can write a general expression for a coordinate transformation in four space-time dimensions as

$$dx^{\beta'} = \frac{\partial x^{\beta'}}{\partial x^{\alpha}} dx^{\alpha}. \tag{2.51}$$

We will use Greek letters for the indices whenever the index runs over all four space-time coordinates. Note that here there is an implied sum over the α index because it is repeated on the right-hand side of the equals sign. The β' index is not summed, because it is not repeated on the same side of the equals sign, but it can take four different values. Hence Equations (2.51) represent four equations, one for each of the four possible values of β', and each equation is, in general, a linear combination of four terms. So, for example, let us set $\beta' = 1$, resulting in

$$dx^{1'} = \frac{\partial x^{1'}}{\partial x^0} dx^0 + \frac{\partial x^{1'}}{\partial x^1} dx^1 + \frac{\partial x^{1'}}{\partial x^2} dx^2 + \frac{\partial x^{1'}}{\partial x^3} dx^3 \tag{2.52}$$

or, in terms of $c\,dt, dx, dy, dz$, and dx',

$$dx' = \frac{\partial x'}{\partial ct} d(ct) + \frac{\partial x'}{\partial x} dx + \frac{\partial x'}{\partial y} dy + \frac{\partial x'}{\partial z} dz. \tag{2.53}$$

Equations (2.51), (2.52), and (2.53) are valid for any transformation of coordinates. For the special case corresponding to Equation (2.50), we can use the relationships between the unprimed and the primed coordinates to work out the partial derivatives in Equation (2.53) in the case of this specific coordinate transformation. We get $\partial x'/\partial(ct) = -\beta\gamma$, $\partial x'/\partial x = \gamma$, $\partial x'/\partial y = 0$, and $\partial x'/\partial z = 0$. If we substitute these partial derivatives into Equation (2.53), then we get the specific Lorentz transformations written down in Equations (2.50).

This shows that the general form given in Equation (2.51) has the most commonly studied Lorentz transformation as a special case. However, firstly, you can show that other Lorentz transformations may also be represented in the form of this equation. Furthermore, Lorentz transformations are only to transform between the frames of reference of different inertial observers. What about the frames of reference of observers who are accelerating? Well, the principle of equivalence states that any such observer is locally equivalent to a freely falling one. The local bit means that the freely falling reference frame will deviate from the accelerating one as you move away from the point where the two are coincident, but the deviations will once again be quadratic (second-order) in the displacements and hence can be neglected locally. In addition, we may choose to make other coordinate transformations, for example, to work in a curvilinear coordinate system. But all such coordinate transformations have the displacement in the new coordinates expressed as a linear superposition of the same displacement in the old coordinates. Therefore these more general coordinate systems are also represented by Equation (2.51).

We therefore call any quantity whose coordinates transform like those of a small displacement between neighbouring space-time points a tensor of rank $\frac{1}{0}$. Another way of looking at this small displacement is that it is a four-dimensional vector – an arrow between two neighbouring points in space-time. Just as for a small spatial displacement on the plane, this arrow is decomposed into components along the directions of basis vectors:

$$d\vec{r} = dx^\alpha \vec{e}_\alpha. \tag{2.54}$$

Just as in the two-dimensional space, the arrow itself is unaffected by different choices of coordinate; therefore the basis vectors along which the components lie must transform like the components of tensors of rank $\frac{0}{1}$. Taking the dot product of $d\vec{r}$ with itself, we find that

$$d\vec{r} \cdot d\vec{r} = dx^\alpha \, dx^\beta \vec{e}_\alpha \cdot \vec{e}_\beta. \tag{2.55}$$

The dot product between $\vec{e}_\alpha$ and $\vec{e}_\beta$ causes the overall product, which is known as the interval between the space-time points, to be a Lorentz invariant. Again following the case of small displacements on the plane, we write

$$g_{\alpha\beta} = \vec{e}_\alpha \cdot \vec{e}_\beta. \tag{2.56}$$

Going back to the special case where both the unprimed and primed observers are inertial, we have $g_{00} = g_{0'0'} = -1$, $g_{11} = g_{1'1'} = g_{22} = g_{2'2'} = g_{33} = g_{3'3'} = +1$, and $g_{\alpha\neq\beta} = g_{\mu'\neq\nu'} = 0$. Therefore we can write

$$\begin{aligned} g_{\alpha\beta} \, dx^\alpha \, dx^\beta &= -c^2 \, dt^2 + dx^2 + dy^2 + dz^2 \\ &= g_{\mu'\nu'} \, dx^{\mu'} \, dx^{\nu'} \\ &= -c^2 \, dt'^2 + dx'^2 + dy'^2 + dz'^2 \\ &= |d\vec{r}|^2 \\ &= ds^2. \end{aligned} \tag{2.57}$$

This reproduces the Lorentz invariance of the interval. However, we should always remember that tensor equations are valid in any coordinate system. Because dx^α and $g_{\alpha\beta}$ are tensors, we also have $g_{\omega''\phi''} \, dx^{\omega''} \, dx^{\phi''} = g_{\alpha\beta} \, dx^\alpha \, dx^\beta$, where the double primed coordinate system is a curvilinear system used to describe the trajectory of an accelerating observer, one who is not in free fall in a gravitational field. Tensors have known transformation properties between general coordinates and between coordinate systems attached to inertial observers, so that a combination of tensor components that yields a Lorentz invariant is also invariant under transformations to more exotic curvilinear space-time coordinate systems.

2.8 Some More Space-Time Tensors

Suppose we have a tensor of rank $\frac{1}{0}$, say its components are A^μ. This means that under a coordinate transformation, it becomes

$$A^{\beta'} = \frac{\partial x^{\beta'}}{\partial x^\mu} A^\mu. \tag{2.58}$$

Notice that the partial derivatives that are the coefficients of the linear combination of A^0, A^1, A^2, and A^3 are the same partial derivatives as they were for the only other tensor of rank $\frac{1}{0}$ whose components we have met so far, dx^μ. This means that, for example, if I have two inertial observers where the primed observer is moving as before at velocity $v = \beta c$ parallel to the x axis, and the x and x' axes are aligned, then the transformation on the components A^μ is

$$A^{0'} = \gamma A^0 - \beta\gamma A^1$$
$$A^{1'} = \gamma A^1 - \beta\gamma A^0$$
$$A^{2'} = A^2$$
$$A^{3'} = A^3. \tag{2.59}$$

What other tensors of rank $\frac{1}{0}$ are common? One example you may already have met is called the 4-momentum. If a particle has total energy E and momentum components in a Cartesian coordinate system of (p^x, p^y, p^z), then the four components of its 4-momentum are $p^0 = E/c$, $p^1 = p^x$, $p^2 = p^y$, $p^3 = p^z$. Indeed, the four components of the 4-momentum transform exactly like the four components dx^α, so that for the two inertial observers previously described, the transformation between the unprimed and primed coordinate systems is

$$\frac{E'}{c} = \gamma\frac{E}{c} - \beta\gamma p^x$$
$$p^{x'} = \gamma p^x - \beta\gamma\frac{E}{c}$$
$$p^{y'} = p^y$$
$$p^{z'} = p^z. \tag{2.60}$$

Recalling that the combination $g_{\alpha\beta}\,dx^\alpha\,dx^\beta$ is the interval ds^2 and is a Lorentz invariant, we can make another Lorentz invariant by combining the metric coefficients with the 4-momentum components, $g_{\alpha\beta}\,p^\alpha\,p^\beta$. In Cartesian coordinates, let us see what this quantity is:

$$g_{\alpha\beta}\,p^\alpha\,p^\beta = -p^0 p^0 + p^1 p^1 + p^2 p^2 + p^3 p^3$$

$$= -\frac{E^2}{c^2} + (p^x)^2 + (p^y)^2 + (p^z)^2$$

$$= -\frac{E^2}{c^2} + |\vec{p}|^2. \tag{2.61}$$

To see what this object is, multiply it by $-c^2$ to get $E^2 - |\vec{p}|^2 c^2$. From the energy–momentum–mass relation we know that this combination is equal to $m_0^2 c^4$, where m_0 is the rest mass of the particle, so $E^2 - p^2 c^2 = m_0^2 c^4$. Hence

$$g_{\alpha\beta} p^\alpha p^\beta = -m_0^2 c^2. \tag{2.62}$$

Again, this is a tensor equation, so it is true for all observers, both inertial and accelerating. We could also form another Lorentz invariant quantity by combining the dx^α with the p^β and the metric coefficients,

$$g_{\alpha\beta} p^\alpha dx^\beta. \tag{2.63}$$

It is not immediately obvious what invariant this is, but it must be one because it is constructed entirely out of tensors, and all the indices are summed over, hence it must be an invariant, otherwise known as a tensor of rank $\genfrac{}{}{0pt}{}{0}{0}$. To figure out what this is, go to a flat space-time with a Cartesian coordinate system for the spatial coordinates and assume that the particle in question is moving along the x axis with momentum of magnitude p. The invariant in this frame is

$$-\frac{E}{c} c\, dt + p\, dx = -E\, dt + p\, dx$$

$$= \hbar(k\, dx - \omega\, dt)$$

$$= \hbar\, d\phi. \tag{2.64}$$

This invariant can be thought of as the phase accumulated by wavefunctions that might represent particles in the quantum limit in a small portion of its world-line multiplied by $\hbar$. Again, because this is a tensor equation, this accumulated phase is something all observers will agree on, no matter whether inertial or accelerating. This is a good thing; otherwise, interference patterns between wave packets would result in different observers finding differing results for the probability densities arising out of quantum theories.

We have identified several useful tensors of rank $\genfrac{}{}{0pt}{}{1}{0}$. What about a tensor of a different rank? Suppose we have some scalar field, perhaps $\Phi(x)$, where x is short-hand for any space-time point, so the field potentially takes a different numerical value at every point in four-dimensional space-time. At any point, there are four partial derivatives you can take of Φ, and they can be written in some unprimed frame as

$$\frac{\partial \Phi}{\partial x^\alpha}. \tag{2.65}$$

A different observer with a different coordinate system will measure different partial derivatives on the same field,

$$\frac{\partial \Phi}{\partial x^{\beta'}}.$$ (2.66)

These two sets of partial derivatives are related using the chain rule of partial differentiation, so that

$$\frac{\partial \Phi}{\partial x^{\beta'}} = \frac{\partial \Phi}{\partial x^{\alpha}} \frac{\partial x^{\alpha}}{\partial x^{\beta'}},$$ (2.67)

where again the repeated α on the right implies a sum. Looking back at Equation (2.27), we see that this is how the components of a tensor of rank $\binom{0}{1}$ transform. This means that the four partial derivatives of a scalar field with respect to ct, x, y, and z, taken together, are the components of a tensor of rank $\binom{0}{1}$. This leads to a common abbreviated form for these partial derivatives,

$$\partial_{\alpha} \Phi = \frac{\partial \Phi}{\partial x^{\alpha}}.$$ (2.68)

With this notation, the transformation law for $\partial_{\alpha} \Phi$ is written as

$$\partial_{\beta'} \Phi = \frac{\partial x^{\alpha}}{\partial x^{\beta'}} \partial_{\alpha} \Phi.$$ (2.69)

2.9 Index Tricks of the Trade

The Einstein summation convention means that long and complicated expressions with many terms can be written very quickly and compactly. Sometimes however if you had the patience to write out all the terms, you would be able to see simplifications and cancellations that are more obscure in the compact form. There are many tricks and rules that can be used to manipulate tensor expressions in ways that are equivalent to operations on the sum of all the terms represented in the Einstein sum, such as re-ordering the terms. Here we go through some of the more useful of these tricks.

2.9.1 Contraction

In Section 2.8, we took the metric coefficients $g_{\alpha\beta}$ and combined them with the 4-displacement dx^{μ} and the 4-momentum p^{ν} to make three different invariant quantities. Let us see what we did, step by step for one of the cases, say, for example, the combination that led to $\hbar \, d\phi$. First, we write a product of tensors,

$$g_{\alpha\beta} \, p^{\nu} \, dx^{\mu}.$$ (2.70)

This is a tensor expression because everything in it is a tensor. The expression as a whole is a tensor of rank $\frac{2}{2}$, because there are two upstairs indices and two downstairs indices. Now we take a pair of indices, one in an upstairs position and the other in a downstairs position, and we set the two indices equal. We do this first with one pair of indices, say here we set α equal to ν, obtaining

$$g_{\nu\beta}\, p^{\nu}\, dx^{\mu}. \tag{2.71}$$

This is still a tensor expression, but it has two less free indices – indices that are not summed over. The rank of this tensor is $\frac{1}{1}$ because it has one free index upstairs and one free index downstairs. Finally, we can if we want to set the remaining free indices equal to each other, and we finally obtain

$$g_{\nu\mu}\, p^{\nu}\, dx^{\mu}. \tag{2.72}$$

This expression now has no free indices, indices that are not summed over, but it is still a tensor expression of rank $\frac{0}{0}$, and therefore it is an invariant.

The process of setting a pair of indices, one upstairs and one downstairs to the same symbol, and therefore effectively summing over the terms where the indices in these two slots are the same is called contraction. Note that you can only contract an upper index with a lower index – if you contract a lower index with another lower index, the resulting sum ceases to be a tensor. However, if you want to do this anyway, then the next sections show you how you can apply contraction to indices initially both in the upper or both in the lower positions, by first using the index raising or lowering procedure.

2.9.2 Index Raising and Lowering

We have already written $ds^2 = g_{\alpha\beta}\, dx^{\alpha}\, dx^{\beta}$. We have so far viewed this expression as the product of three tensors, one of rank $\frac{0}{2}$ and two more of rank $\frac{1}{0}$. However, let us consider another way of viewing this product by reordering it and inserting some brackets:

$$
\begin{aligned}
ds^2 &= g_{\alpha\beta}\, dx^{\alpha}\, dx^{\beta} \\
     &= \left(dx^{\alpha}\, g_{\alpha\beta}\right) dx^{\beta}.
\end{aligned} \tag{2.73}
$$

The product in brackets is a tensor of rank $\frac{0}{1}$ because it has one free lower index. It contains the metric coefficients, which contain information about the local properties of the coordinate system and beyond that the same set of displacement components present in dx^{β}. Let us therefore write a new symbol for the product of $g_{\alpha\beta}\, dx^{\alpha}$:

$$dx_{\beta} = g_{\alpha\beta}\, dx^{\alpha}, \tag{2.74}$$

where, once again, the repeated α index implies a sum. Following exactly the same argument, we can also write

$$p_\mu = g_{\mu\nu} p^\nu. \tag{2.75}$$

It appears that we can consider the metric coefficients $g_{\alpha\beta}$ as an index lowering operator. The objects dx_β and p_μ contain the same information as their counterparts dx^α and p^ν, but whereas the latter transform as the components of tensors of rank $\begin{smallmatrix}1\\0\end{smallmatrix}$, the former transform as the components of tensors of rank $\begin{smallmatrix}0\\1\end{smallmatrix}$.

What about raising indices? Let us call our index-raising operator $g^{\alpha\mu}$ without yet knowing what it is. Since it is an index-raising operator, it must be able to convert our newly minted p_μ back into an index-raised object. So, for example, we must be able to write

$$g^{\alpha\mu} p_\mu = p^\alpha. \tag{2.76}$$

We insert Equation (2.76) into Equation (2.75) to obtain

$$p^\alpha = g^{\alpha\mu} g_{\mu\nu} p^\nu, \tag{2.77}$$

so that p^α is the result of starting with p^ν transforming to the equivalent lowered-index tensor and then transforming back again. The only way this can work is if

$$g^{\alpha\mu} g_{\mu\nu} = \delta^\alpha_\nu, \tag{2.78}$$

where δ^α_ν is the four-dimensional Kronecker delta, equal to one when $\alpha = \nu$ and zero otherwise. Put another way, if you write the elements of $g^{\alpha\mu}$ in a square 4×4 matrix, and the elements of $g_{\beta\nu}$ in another square matrix, then multiple the two matrices together using the normal rules of matrix notation, which incidentally are exactly the same as contracting the μ index with the β index in $g^{\alpha\mu} g_{\beta\nu}$, then the resultant 4×4 matrix is the identity matrix. So, the elements of $g^{\alpha\mu}$, which we will call the inverse metric, can be obtained from the elements of $g_{\beta\nu}$ by matrix inversion. Of course, in general, inverting a 4×4 matrix is not easy, but fortunately in general relativity, there are many metrics that make very sparse matrices (matrices with a lot of zeros in), and so it is often reasonably easy to compute the inverse of metric matrices. In Section 1.8, I gave you some examples of this in the diagnostic test for practice.

Contraction also gives us a shorter hand way of writing some of our favourite invariants. You can now write the squared length of a short space-time displacement $|d\vec{r}|^2$ as

$$|d\vec{r}|^2 = dx^\mu \, dx_\mu, \tag{2.79}$$

and the energy momentum mass relation can be written as

$$p^\alpha p_\alpha = -m_0^2 c^2. \tag{2.80}$$

The factors of the metric coefficients are still there, but they are hidden as the lowering operators that were used to turn p^α into p_β, and so forth.

Tensors of higher rank can have each of their indices raised or lowered using appropriate factors of $g_{\mu\nu}$ and $g^{\mu\nu}$. For example, if $R^\alpha{}_{\beta\gamma\delta}$ are the components of a tensor of rank $\frac{1}{3}$, then we could raise the second index like this:

$$R^{\alpha\phi}{}_{\gamma\delta} = g^{\phi\beta} R^\alpha{}_{\beta\gamma\delta}. \tag{2.81}$$

This highlights an annoyance of tensor notation, which is that the order of indices matters even when a mixture of upstairs and downstairs indices is present. You can see this by considering an example – an arbitrary tensor of rank $\frac{2}{0}$ with components $F^{\alpha\beta}$. This tensor is neither symmetric nor antisymmetric in the interchange of its two indices. We lower one of the indices with $g_{\beta\gamma}$, resulting in a tensor of rank $\frac{1}{1}$, which we might sloppily call F^α_γ, but then if we decide to raise the γ index using $g^{\delta\gamma}$, then we might forget that the γ index originated in the index β that was very definitely the second of the two indices in the original tensor, and end up with $F^{\delta\alpha}$. So we have ended up with a different tensor from the original one by performing an operation that is supposed to get us back to the tensor we started with.

To avoid this problem, our intermediate tensor of rank $\frac{1}{1}$ is more properly written $F^\alpha{}_\gamma$. The two indices have a well-defined order despite the fact that one of them is upstairs and the other is downstairs. You just have to write them neatly enough so you can discern the order of the indices from the gaps.

2.9.3 Swapping Summed Indices

Going back to the expression $g_{\alpha\beta}\, dx^\alpha\, p^\beta$, there are implied sums over both α and β so that in general this expression represents 16 terms. Because of the implied sums, the indices α and β can be swapped without changing the value of the expression, so that

$$g_{\alpha\beta}\, dx^\alpha\, p^\beta = g_{\beta\alpha}\, dx^\beta\, p^\alpha. \tag{2.82}$$

In addition, if you have two terms added together both containing the same dummy indices, it is perfectly fine to swap the indices in one of the terms but not the other one. So, for example, you can do this:

$$g_{\alpha\beta}\, dx^\alpha\, p^\beta - g_{\alpha\beta}\, dx^\beta\, q^\alpha = g_{\alpha\beta}\, dx^\alpha\, p^\beta - g_{\beta\alpha}\, dx^\alpha\, q^\beta$$
$$= dx^\alpha \left(g_{\alpha\beta} p^\beta - g_{\beta\alpha} q^\beta \right). \tag{2.83}$$

This can be useful when you want to take a common factor out of two terms, but the indices in one of the terms do not initially match the indices in the other term.

2.9.4 Symmetry and Antisymmetry

A tensor is symmetric in two of its indices if interchanging the two indices leaves the tensor components the same. We have already met one symmetric rank $\binom{0}{2}$ tensor, because $g_{\alpha\beta} = \vec{e}_\alpha \cdot \vec{e}_\beta$, so that $g_{\beta\alpha} = g_{\alpha\beta}$ because dot products are commutative. A tensor is antisymmetric with respect to two of its indices if interchanging them results in a sign change.

If you have a general tensor with more than one index either upstairs or downstairs, say, for example, $M^{\alpha\beta}$, then you can construct symmetric and antisymmetric tensors $M_S^{\alpha\beta}$ and $M_A^{\alpha\beta}$ out of this general one as follows:

$$M_S^{\alpha\beta} = M^{\alpha\beta} + M^{\beta\alpha}$$
$$M_A^{\alpha\beta} = M^{\alpha\beta} - M^{\beta\alpha}. \tag{2.84}$$

In both cases, interchanging α and β swaps the two terms, and in the upper case, this results in no change because both terms have the same sign, whereas in the lower case a sign change results because of the relative minus sign between the two terms. Using this trick, you can decompose any tensor having a pair of indices either upstairs or downstairs into a symmetric and antisymmetric component, because from Equations (2.84) we have

$$M^{\alpha\beta} = \frac{1}{2}\left(M_S^{\alpha\beta} + M_A^{\alpha\beta}\right). \tag{2.85}$$

This sometimes is called symmetrisation and antisymmetrisation of the tensor. Other objects, such as functions, can be symmetrized and antisymmetrized with respect to their arguments.

Symmetric and antisymmetric tensors have special properties when they are contracted with each other. For example, if you fully contract a symmetric tensor with an antisymmetric one, then the result is zero. For example,

$$M_S^{\alpha\beta} M_{A\alpha\beta} = 0, \tag{2.86}$$

where I suppose that M_A with lower indices must have had its indices lowered using two factors of the index lowering operator $g_{\mu\nu}$:

$$M_{A\alpha\beta} = g_{\alpha\mu}g_{\beta\nu}M_A^{\mu\nu}. \tag{2.87}$$

2.10 Problems

2.1 A portion of the two-dimensional plane can be described in plane hyperbolic coordinates (ε, η), where

$$x = \varepsilon \cosh \eta$$
$$y = \varepsilon \sinh \eta,$$

and η and ε are real numbers.

(a) Sketch and describe the lines of constant ε.
(b) Sketch and describe the lines of constant η.
(c) What portion of the XY plane is not covered in plane hyperbolic coordinates?
(d) Work out expressions for the basis vectors $\vec{e}_\eta$ and $\vec{e}_\varepsilon$ in terms of the coordinates ε, η, $\vec{e}_x$, and $\vec{e}_y$.
(e) Work out the components of the metric in plane hyperbolic coordinates in terms of η and ε.
(f) Work out an expression for the transformation between the plane hyperbolic components $(d\varepsilon, d\eta)$ of a small displacement $d\vec{r}$ in the portion of the plane covered by the plane hyperbolic coordinate system to the same small displacement in terms of the Cartesian components of the displacement (dx, dy).
(g) Show that $d\vec{r} = dx\,\vec{e}_x + dy\,\vec{e}_y = d\eta\,\vec{e}_\eta + d\varepsilon\,\vec{e}_\varepsilon$. Use explicit substitution as was carried out in Section 2.3 for the case of plane polar coordinates, rather than by asserting the general results.

2.2 The conversion from Cartesian to spherical polar coordinates takes place by the following substitutions: $x = r \sin\theta \cos\phi$, $y = r \sin\theta \cos\phi$, and $z = r\cos\theta$, where θ is the so-called colatitude (the latitude θ^{L} is by convention the angle of elevation above the equator, $\theta^{\mathrm{L}} = \pi - \theta$), and ϕ is the longitude (in navigation) or the azimuth (in astronomy and geometry).

(a) Find expressions for the basis vectors $\vec{e}_r$, $\vec{e}_\theta$, and $\vec{e}_\phi$ in terms of r, θ, ϕ, $\vec{e}_x$, $\vec{e}_y$, and $\vec{e}_z$.
(b) Find expressions for the coefficients of the metric in spherical polar coordinates.
(c) Find expressions for the coefficients g^{ij} of the inverse metric.

2.3 Consider a spherical surface, the set of all points at the same radius R from the origin. In terms of spherical coordinates a point on the surface can be specified by coordinates (θ, ϕ), which are the angle coordinates of the spherical coordinate system. You do not need R to be a coordinate because R is fixed. Work out the components of the metric in this coordinate system. Hint: because there are only two coordinates, the metric has three independent components. However, you will need to start by describing points on the spherical surface in terms of their three Cartesian components to arrive at the result.

2.4   Consider a function $f(x, y)$ of two variables x and y. In turn, x and y are both functions of two further variables u and v, $x(u, v)$ and $y(u, v)$, By considering the expression for the total derivative of a function of two variables from the notes, show that

$$\left(\frac{\partial f}{\partial u}\right)_v = \left(\frac{\partial f}{\partial x}\right)_y \left(\frac{\partial x}{\partial u}\right)_v + \left(\frac{\partial f}{\partial y}\right)_x \left(\frac{\partial y}{\partial u}\right)_v .$$

In the case where $x(u, v)$ and $y(u, v)$ are invertible, so that $u(x, y)$ and $v(x, y)$ are also well-defined functions, show in addition that

$$\left(\frac{\partial f}{\partial y}\right)_x = \left(\frac{\partial f}{\partial u}\right)_v \left(\frac{\partial u}{\partial y}\right)_x + \left(\frac{\partial f}{\partial v}\right)_u \left(\frac{\partial v}{\partial y}\right)_x .$$

2.5   By equating the Kronecker delta δ_l^k with the partial derivative $\partial x^k / \partial x^l$ and using the chain rule for partial derivatives discussed in Problem 2.4, show that the components of δ_l^k transform as the components of a tensor of rank $\frac{1}{1}$.

2.6   In Section 2.2, we saw that our basis vectors defined by $\vec{e}_i = \partial \vec{r} / \partial x^i$ are not generally of unit length. Show that a unit vector $\hat{e}_i$ and the basis vector $\vec{e}_i$ in the same direction as previously defined are related by (no Einstein sum over index j here) $\vec{e}_j = \sqrt{g_{jj}} \hat{e}_j$ or, equivalently, $\hat{e}_j = \vec{e}_j / \sqrt{g^{jj}}$.

2.7   In Section 2.9.4, we discussed creating symmetric and antisymmetric objects under interchange of indices. Consider a tensor of rank $\frac{0}{3}$ in three spatial dimensions with no particular symmetry under interchange of indices, whose components are a_{ijk}. You can construct a tensor whose components are antisymmetric under the interchange of the first two indices, $P_{ijk} = a_{ijk} - a_{jik}$.

   (a)   Repeat the procedure twice more with the remaining pairs of indices to obtain a tensor Q_{ijk} that is antisymmetric with respect to the interchange of any pair of indices.
   (b)   Are Q_{ijk} the components of a tensor? If so, why? If not, why not?
   (c)   Show that Q_{ijk} is symmetric with respect to cyclic permutation of the indices, that is, $Q_{ijk} = Q_{jki} = Q_{kij}$.
   (d)   Show that components of Q_{ijk} where any two or more of the indices are the same are zero.
   (e)   Specialising to the case where Q_{ijk} consists only of constants, that is, it does not have functional dependence on any variable such as position or time, show that there are only six non-zero components and they are all determined by a single numerical parameter.

2.8   In Problem 2.7, we showed that a three-index object antisymmetric under interchange of any pair of its indices and independent of any parameter, in particular, the same at any point in space, is very restricted in the values of its components. In fact, there is only one overall freedom, a numerical scale factor to all the elements, which we just set equal to 1. Then by convention we define the components of the Levi-Civita fully antisymmetric tensor ε_{ijk}. It is antisymmetric under interchange of any index pair, and by convention $\varepsilon_{123} = +1$. This tensor is important enough to be named after a famous Italian mathematician, the Levi-Civita tensor. Tullio Levi-Civita was one of the founders of differential geometry, before Einstein adopted the theory to formulate general relativity.

(a) Work out the values of all components of ε_{ijk}.

(b) Using the result from (a), write an expression for ε_{ijk} in terms a sum of products of Kronecker deltas. For example, a term $\delta_i^1 \delta_j^2 \delta_k^3$ could be used to make $\varepsilon_{123} = +1$. Repeat the argument for the other non-zero components.

(c) In Problem 2.5, we showed that the Kronecker delta components δ_j^i are components of a tensor of rank $\frac{1}{1}$. Show further that a partially restricted Kronecker delta, for example δ_j^3, where the upper index is constrained to take a particular value, 3 in this case, are the components of a tensor of rank $\frac{0}{1}$. As a consequence of this part and part (b), you have shown that ε_{ijk} transforms as a tensor of rank $\frac{0}{3}$.

2.9   The Levi-Civita tensor is actually strictly referred to as a pseudo-tensor. This is because when you make scalars using the Levi-Civita tensor components, they do not behave quite like the scalars we have met before under the discrete operation called parity. Parity consists of making the discrete transformation $\vec{r} \to -\vec{r}$. Show that if we make a scalar $a = \varepsilon_{ijk} x^i x^j x^k$, then under parity, $a \to -a$. Therefore a is called a pseudo-scalar, because it transforms under coordinate transformations like a tensor of rank $\frac{0}{0}$, but under the discrete parity transformation, it picks up a minus sign. By contrast, an example of a scalar is the simple product $s = x^i x_i$, Show that this product does not change sign under parity.

2.10 In Cartesian coordinates, starting with $\varepsilon_{123} = 1$, show that $\varepsilon^{123} = 1$. Using the general rule for the transformation law for the components of a tensor of arbitrary rank given in Equation (2.32), write down the transformation law to the same tensor in a different coordinate system, $\varepsilon^{l'm'n'}$. What is $\varepsilon^{r\theta\phi}$ in spherical polar coordinates defined in Problem 2.2? Hint: it is not $+1$. You will need to take the partial derivatives of r, θ, and ϕ with respect to x, y, and z very carefully to get the right answer. Though there are lots of terms, at the end, there is also a lot of cancellation.

3

Matter in Space-Time

We now turn to the physical contents of space-time. We said in Chapter 1 that the cause of curvature in space-time is the presence of matter and energy. In special relativity, you learned that a particle of rest mass m_0 moving at velocity $v = \beta c$ has the total energy $\gamma m_0 c^2$, where $\gamma = 1/\sqrt{1 - \beta^2}$. What we are faced with in our Universe, however, is something more interesting than a mere single particle. We are faced with a diversity of different constituents from ordinary 'baryonic' matter to highly relativistic electromagnetic radiation, neutrinos, and cosmic rays to cold dark matter, so-called dark energy, and even the energy content of space-time curvature itself.

A full examination of these constituents is beyond the scope of this book, and indeed some of them remain mysterious. In this chapter, we are going to work out how to represent the simplest constituent we could imagine, called dust, using a new tensor. We will work in ordinary flat space-time, where we have at least some intuition. What we are after is an expression for the energy and matter content of space-time arising from this dust. The advantage of this approach is that as a tensor, this entity will then have known transformation properties to the reference frames of all other observers, enabling us to use our tensor in the curved space-time that accelerating observers observe as a gravitational field. We will also assert that some of the properties of the tensor describing our dust are also properties of the analogous tensors for other constituents of the Universe, so that we can seek these same properties in other tensors that we will be led to in Chapter 4.

3.1 Dust

Dust is a term that in the context of cosmology means particles all of the same rest mass m_0 that are stationary in some coordinate system. Starting in that coordinate system, assume that we have some dust and that over a cubic unit volume of space, the dust has a fixed number density n. Because in this rest frame, the dust is at rest, no particles pass through the walls. Now imagine that we boost to some reference frame moving at velocity v^x parallel to the x axis of the rest frame. In this reference frame, several things happen. The first thing that happens is that there is Lorentz contraction parallel to the direction of motion of this reference frame. This causes all the particles of dust to compress parallel to this direction, so that the number density is increased by a factor of γ. Then the number density becomes

$$n' = \frac{n}{\sqrt{1-\beta^2}}. \tag{3.1}$$

The next thing that happens is that the primed observer can measure a flux of particles through a wall perpendicular to the motion of their frame. Again, let this wall be of unit cross sectional area. In a time of 1 second in this moving frame, this wall sweeps out a volume of $1\,\mathrm{m} \times 1\,\mathrm{m} \times v^x\,\mathrm{m\,s}^{-1} \times 1\,\mathrm{s} = v^x\,\mathrm{m}^3$. Therefore the number of particles that pass through the unit area end wall in this time is the number density $\overline{n'}$ multiplied by the volume swept out. The number of particles passing through a unit area in unit time is called a flux. So, the flux of dust particles through a surface normal to the direction of motion is

$$\Phi^{x'} = \frac{n v^{x'}}{\sqrt{1-\beta^2}}. \tag{3.2}$$

Suppose now that the velocity of our primed frame is actually in some arbitrary direction and therefore has components (v^x, v^y, v^z). This means that particles will pass through $1\,\mathrm{m}^2$ areas normal to all three axes at a rate given by the velocity components in those three directions. We then have three copies of Equation (3.2), one for each of the three position directions, which express how many particles from the dust pass through a unit area aligned normal to each of the three axes in the primed frame. These equations can be expressed collectively as the components of a vector flux through surfaces normal to the three Cartesian axes.

$$\Phi^{i'} = \frac{n v^{i'}}{\sqrt{1-\beta^2}}. \tag{3.3}$$

Now we are aware that though we have used an index i in Equation (3.3) to denote the three spatial directions, we are going to be wanting indices that run over all four directions. Do we have a fourth component for this object? An obvious candidate is the number density given in Equation (3.1). However, firstly, it has different dimensions from the components of flux, and, secondly, it is not obviously a flux.

In fact, we can expand the three-dimensional concept of a flux into space-time. The key is to consider the direction corresponding to the flux measurement. For example, measurement of the x component of flux was conducted by counting particles flowing parallel to the x axis, and similarly for y and z. For the fourth component, then perhaps we should consider the flux of particles flowing parallel to the t axis. This is simplest to do if you consider stationary particles, though just as when we first worked out the x component of flux, we neglected motion in the y or z directions, it will generalise naturally. On a space-time diagram. particles that are stationary nevertheless follow world lines. Where the world lines represent stationary particles, they run parallel to the ct axis. Stationary particles move along world lines parallel the time axis, just as when we boost from the frame in which dust is at rest to one where the observer is moving in some direction, the dust particles move parallel to their direction of motion.

The next question is how fast do the particles move along the world lines. Imagine for a second that the ct axis of the space-time diagram was just time t in seconds. The rate at which the point, or event, momentarily coincident with the particle moves up the world line would be 1 unit every second. Now, on a space-time diagram, we in fact use the coordinate ct instead of just t. In these units, then, stationary particles move up the world lines at a rate of c units per second. Following exactly the same procedure as we did for the spatial components of the flux, we multiply the velocity along the world line by the number density of the dust. Recall that we are in a frame where the observer is moving with respect to the dust in the spatial directions as well, so that the number density we need for the dust is the number density in the primed frame from Equation (3.1). So, we finally arrive at an expression for the zeroth component of the flux,

$$\Phi^{0'} = \frac{nc}{\sqrt{1 - \beta^2}}. \tag{3.4}$$

Do Equations (3.4) and (3.3) group together into a 4-vector equation for the 4-flux? Yes, they do, but you might have forgotten the four vector we need. Recall that the four-momentum of a particle has components $p^0 = E/c = \gamma m_0 c$ and $p^i = \gamma m_0 v^i$. We can interpret this as the rest mass m_0 multiplied by a four vector called the four-velocity, whose four components are

$$U^0 = \gamma c \qquad\qquad U^i = \gamma v^i. \tag{3.5}$$

In terms of the four-velocity of the dust, the four-flux of particles for the primed observer is

$$\Phi^{\alpha'} = n U^{\alpha'}, \tag{3.6}$$

where the primes are there to remind us that our observer is in an inertial frame at motion with respect to the dust particles.

3.2 Energy and Momentum of Dust

We next imagine that we are using dust as a source for the gravitational field. In fact, in parts of the Universe, it is indeed true that dust is the dominant source for gravitational fields, so this is not inappropriate. Though we have identified a tensor to express quantities like the number density and the flux of particles through a surface, we are not there yet, because this is not telling us the energy content of the dust in any way; it is telling us how much dust there is and how much is flowing in different directions. If we want to introduce energy into the mix, then we need to consider the energy of the dust particles.

Start by considering the dust at rest. In a metre-cubed volume of dust, there are n particles of mass m_0, so the energy of this dust is nm_0c^2. Now, however, consider what happens when the dust starts moving. There are two relevant effects. First, the density of the dust increases because of the Lorentz contraction, looking back to Equation (3.1). Second, each individual dust particle gains energy due to its motion, so that its total energy is now

$$ E = \frac{m_0c^2}{\sqrt{1 - \beta^2}}. \tag{3.7} $$

Consequently, the energy density in a frame in which the dust is moving is given by the modified number density times the modified energy per particle, or

$$
\begin{aligned}
\varepsilon' &= \frac{n}{\sqrt{1 - \beta^2}} \times \frac{m_0c^2}{\sqrt{1 - \beta^2}} \\
&= \frac{nm_0c^2}{1 - \beta^2} \\
&= \frac{\varepsilon}{1 - \beta^2},
\end{aligned}
\tag{3.8}
$$

where ε is the energy density of the dust in the frame where it is at rest.

3.3 The Stress Energy Tensor of Dust

We know, however, that this cannot be the end of the story. When the dust moves, its particles travel from place to place, carrying energy and momentum with them. We have the four-momentum of each particle that yields the four components of the energy and momentum, and the four-flux, which yields the four components of the flux in space-time. Let us combine these two objects in the simplest way that we can by defining

$$ T^{\mu'\nu'} = p^{\mu'} \Phi^{\nu'}, \tag{3.9} $$

and see whether this combination corresponds to the physical picture we have built up. Consider the component with $\mu' = \nu' = 0$,

$$
\begin{aligned}
T^{0'0'} = p^{0'} \Phi^{0'} &= \frac{E'}{c} \times \frac{nc}{\sqrt{1 - \beta^2}} \\
&= \frac{m_0 c}{\sqrt{1 - \beta^2}} \frac{nc}{\sqrt{1 - \beta^2}} \\
&= \frac{nm_0 c^2}{1 - \beta^2} \\
&= \frac{\varepsilon}{1 - \beta^2},
\end{aligned} \tag{3.10}
$$

which is the energy density in the primed frame, calculated in Equation (3.8). Next, let us set $\mu' = 0'$ and $\nu' = i'$, and evaluate that component. This means we are combining the energy density with the 0'th component of the flux, which is going to yield the something we can physically interpret as the i' component of the energy flux, divided by a factor of c for dimensional consistency with the $T^{0'0'}$ component,

$$
\begin{aligned}
T^{0'i'} &= \frac{m_0 c}{\sqrt{1 - \beta^2}} \times \frac{n \nu^{i'}}{\sqrt{1 - \beta^2}} \\
&= \frac{nmc^2 (\nu^{i'}/c)}{1 - \beta^2} \\
&= \frac{\varepsilon (\nu^{i'}/c)}{1 - \beta^2}.
\end{aligned} \tag{3.11}
$$

The picture you need here is that if the box is moving relative to the dust particles, say such that two opposite walls of the box are normal to the direction of motion, then particles are passing into the box through one of these 'end' walls, whilst also passing out through the other end. Thus there is a flux of energy into the box through one wall and out of the box through the opposite wall at the same time.

Next, we take $\mu' = j'$ and $\nu' = 0'$:

$$
\begin{aligned}
T^{j'0'} &= \frac{n \nu^{j'}}{\sqrt{1 - \beta^2}} \times \frac{m_0 c}{\sqrt{1 - \beta^2}} \\
&= \frac{nm_0 c^2 (\nu^{j'}/c)}{1 - \beta^2} \\
&= \frac{\varepsilon (\nu^{j'}/c)}{1 - \beta^2}.
\end{aligned} \tag{3.12}
$$

Here we imagine the dust in the boxes interior when the box is moving, again, in the direction normal to two of the walls. Just as dust is flowing in through one wall and out through the opposite wall, dust in the interior of the box is also flowing from one wall to the other, and therefore the interior of the box contains dust with momentum. Dividing the sum of the momenta of all the dust particles contained by the volume of the box, we see that this component is the 'momentum density' of the dust in the box, divided by c. The momentum density inside the box is numerically equal to the energy flux through the walls normal to the motion.

Finally, we take $\mu' = i'$ and $\nu' = j'$. By analogy with the other components we call these components the momentum flux. There are nine of these components,

$$
\begin{aligned}
T^{i'j'} &= \frac{m_0 v^{i'}}{\sqrt{1-\beta^2}} \times \frac{n v^{j'}}{\sqrt{1-\beta^2}} \\
&= \frac{n m_0 c^2 (v^{i'}/c)(v^{j'}/c)}{1-\beta^2} \\
&= \frac{\varepsilon (v^{i'}/c)(v^{j'}/c)}{1-\beta^2}.
\end{aligned}
\tag{3.13}
$$

To interpret these components physically, we first reason verbally about what the other components might mean. So far we have found that $T^{0'0'}$ represents the energy density, the energy per unit volume in the box. Next, we found that $T^{0'i'}$, the contents of the 0th row and the ith column represent, up to a factor of c the i-component of the energy flux, the rate at which energy flows in through the box walls normal to the three axes. The $T^{j'0'}$ components represent, up to the same factor of c, the momentum density, the three components of the momentum per unit volume inside the box. Algebraically, the three components of the momentum density are equal to the three components of the momentum flux, so at least amongst the elements of the 0th row and column, the tensor is symmetric.

Moving on to the remaining elements $T^{i'j'}$, we might expect that these components can be interpreted as the momentum flux. Furthermore, we might expect that amongst these nine elements we might again find symmetry with respect to $i' \leftrightarrow j'$. These guesses turn out to be correct. To see this, consider Figure 3.1

Special relativistic effects only introduce multiplicative factors $1/(1-\beta^2)$ in all the components of the tensor, as we have already discussed. We therefore neglect these corrections in the following physical argument. Consider the typical particle to the top left of the box, close to the left-hand wall normal to the x axis. This particle is about to exit the box through this wall. It has two non-zero momentum components, one normal to the wall and the other directed along the wall. The flux through the wall is $n v_x$, where v_x is the x-component of the box velocity resolved parallel to the x axis. Notice that this flux drops if the velocity of the box rotates to be more in the y direction. You can think of this as the x end walls of the box presenting a smaller cross section to the flow of particles as they move at larger angles to the walls. The two non-zero momentum components are directed in the

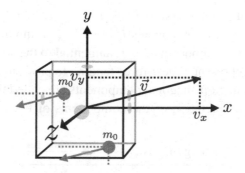

Figure 3.1 Dust particle movements with respect to a box having a velocity $\vec{v}$ in the XY plane. In the rest frame of the box, particles inside it all have the same momentum directed in the opposite direction relative to the velocity vector of the box. These negative momenta have non-zero components in the x and y directions. Those sufficiently near to the walls of the box exit through these walls, leading to a flux of momentum through the walls. Particles exiting through a given wall are replaced by particles having the same momentum entering through the opposite wall, which leads to the divergence-free quality of the stress energy tensor discussed in Section 3.4.

x and y directions and up to relativistic corrections are $-m_0 v^x$ and $-m_0 v^y$. So, two contributions to the negative momentum of particles currently in the box are about to leave it. For the component parallel to the x axis, both the flux and the momentum component contain v^x, so the diagonal components of the momentum flux are proportional to $(v^x)^2$. For the component tangential to the x axis, the momentum contains a v^y and the flux contains a v_x, so the off-diagonal component of the momentum flux corresponding to the flow of momentum y component through the wall normal to the x axis, $T^{2'1'}$, is proportional to $mv^x v^y$.

Now consider the other particle drawn in the bottom right of the box. It is about to leave through the wall normal to the y axis. It also has non-zero momentum components in the x and y directions. It is the momentum component parallel to the x axis, and tangential to the y wall is larger than the tangential momentum of the other particle, but the flux of these particles through their wall, normal to the y axis, is smaller than the flux of particles through the walls normal to the x axis, as the velocity vector is here dominated by the v^x component. So the greater momentum component is compensated for by the smaller flux, and $T^{1'2'}$ is also proportional to $mv^x v^y$. This validates our guess that in the purely spatial components, $T^{i'j'}$ are also symmetric, so that the full set of tensor components $T^{\mu'\nu'}$ is symmetric with respect to the exchange of its indices.

Finally, note that the three diagonal components $T^{i'i'}$ correspond to the momentum flux normal to each of the side walls of the box. The rate at which momentum is transferred to a wall normal to that wall would, were the wall actually to be a physical barrier, be called the pressure on the wall. Accordingly, these on-diagonal components are the pressure components $P^{i'} = T^{i'i'}$ (no sum over i' implied in this equation). Similarly, the components $T^{i'\neq j'}$ tangential to the walls would, were the walls to be physical, represent a sheer force tending to push sideways on each wall, so that therefore the three off-diagonal components are called the sheer stress.

In summary, the interpretation of the components of the energy–momentum–stress tensor is

$$
\left(T^{\mu'\nu'}\right) = \left(
\begin{array}{c|ccc}
\text{Energy Density} & \multicolumn{3}{c}{\dfrac{\text{Energy flux}}{c}} \\
\hline
\dfrac{\text{Momentum density}}{c} & P^{x'} & S^{x'y'} & S^{x'z'} \\
 & S^{x'y'} & P^{y'} & S^{y'z'} \\
 & S^{x'z'} & S^{y'z'} & P^{z'}
\end{array}
\right),
\qquad (3.14)
$$

where $P^{x'}$, $P^{y'}$, and $P^{z'}$ denote the pressure normal to the walls on the three axes, corresponding to outflow of particles normal to the walls, and $S^{x'y'}$, $S^{x'z'}$, and $S^{y'z'}$ are the three components of the sheer stress at the walls.

3.4 Conservation of Energy

We have emphasised throughout this discussion that the flux of energy and momentum in the dust through three mutually perpendicular box walls is balanced by the flux of the same quantities out through the opposite walls. This will lead to the critical observation that the stress energy tensor is divergence-free, a property of this tensor that carries over to more complicated substances than the dust discussed in his chapter. In this section, we work out the mathematics corresponding to this statement.

Let us figure out the consequences of conservation of energy for the stress energy tensor. We drop the primes, so the unprimed frame is now moving with respect to our dust. The energy density is T^{00}. The energy in a volume l^3 is therefore $l^3 T^{00}$. If this quantity changes in time, then there must be a corresponding flow of energy out of the volume. This energy flow can be due to net energy flux in the x, y, or z directions. In the x direction, the net energy flux is the energy flowing through the right-hand wall per second, which is $cl^2 T^{0x}(x = l)$ minus the energy flowing through the left-hand wall per second, which is $cl^2 T^{0x}(x = 0)$. A similar argument is made for the walls normal to the x and y axes. The time rate of change of energy in the box must equal the rate at which energy exits through the ends, so we arrive at

$$
\begin{aligned}
\frac{\partial}{\partial t}\left(l^3 T^{00}\right) = -cl^2 \big[& T^{0x}(x = l) - T^{0x}(x = 0) \\
& + T^{0y}(y = l) - T^{0y}(y = 0) \\
& + T^{0z}(z = l) - T^{0z}(z = 0) \big].
\end{aligned}
\qquad (3.15)
$$

Dividing through by the l^3 and taking the limit as $l \to 0$, we identify the difference between the spatial derivatives in each of the three directions divided by the separation of the planes l on either side of the box, which becomes the partial derivative of the component T^{0i} with respect to x^i. We therefore arrive at

$$\frac{1}{c} \frac{\partial T^{00}}{\partial t} = -\frac{\partial T^{0x}}{\partial x} - \frac{\partial T^{0y}}{\partial y} - \frac{\partial T^{0z}}{\partial z} \tag{3.16}$$

or

$$\frac{\partial T^{0\nu}}{\partial x^\nu} = 0. \tag{3.17}$$

The same argument can be applied to each component of the momentum. The flux of each momentum component through the side walls has to be equal to the time rate of change of the density of that momentum component in the box, and hence we also have

$$\frac{\partial T^{i\nu}}{\partial x^\nu} = 0. \tag{3.18}$$

Putting together Equations (3.17) and (3.18), we conclude that

$$\frac{\partial T^{\mu\nu}}{\partial x^\nu} = 0. \tag{3.19}$$

This is a fundamental property of the stress energy tensor of dust. It turns out to be a fundamental property of the stress energy of other components of the Universe, such as electromagnetic radiation. If the stress energy tensor is to be the source for the curvature of space-time in Einstein's equations, then the tensor describing the curvature itself must share this property, as well as the other property we have found, the symmetry with respect to the exchange of its two indices. These required properties turn out to be decisive factors in the selection from the various tensors that describe curvature of the one that will be inserted into Einsteins equations. Working out suitable tensors for describing curvature is the aim of the next chapter.

3.5 Problems

3.1 Write down the transformation rule for the components of the stress energy tensor $T^{\mu\nu}$ into some arbitrary primed coordinate system.

3.2 Write down the 16 components of the stress energy tensor of dust in its own rest frame. By applying the appropriate transformation to the coordinate system of an inertial observer moving with respect to the dust, arrive at the expression for the components of $T^{\mu'\nu'}$ in the frame with respect to which the dust is in motion.

3.3 The metric for a comoving coordinate system (ct', x', y', z') with Cartesian spatial coordinates in a flat homogeneous isotropic Universe is

$$ds^2 = -c^2 \, dt'^2 + a(t')^2 (dx'^2 + dy'^2 + dz'^2),$$

where $a(t')$ is the scale factor.

(a) Give expressions for the components $c \, dt'$, dx', dy', and dz' in terms of the components $c \, dt$, dx, dy, and dz in a non-comoving Cartesian coordinate system where $ds^2 = -c^2 \, dt^2 + dx^2 + dy^2 + dz^2$.

(b) Give expressions for the comoving basis vectors $\vec{e}_{t'}$, $\vec{e}_{x'}$, $\vec{e}_{y'}$, and $\vec{e}_{z'}$ in terms of the basis vectors $\vec{e}_t$, $\vec{e}_x$, $\vec{e}_y$, and $\vec{e}_z$ in a non-comoving Cartesian coordinate system.

(c) Give expressions for the four non-zero partial derivatives $\partial x^{\alpha'}/\partial x^{\mu}$ between the primed, comoving coordinate system and the unprimed, static Cartesian one. Ensure you specify the values of α' and μ in each case.

An observer determines the components of the energy–momentum–stress tensor for a representative unit volume in this universe, using ordinary laboratory coordinates, and finds that they are

$$\left(T^{\mu\nu}\right) = \begin{pmatrix} \varepsilon & 0 & 0 & 0 \\ 0 & w\varepsilon & 0 & 0 \\ 0 & 0 & w\varepsilon & 0 \\ 0 & 0 & 0 & w\varepsilon \end{pmatrix},$$

where w is a dimensionless number.

(d) Write down the transformation law for converting the components of the energy–stress tensor from laboratory coordinates to comoving coordinates.

(e) Find and write down the components $T^{\alpha'\beta'}$ of the stress tensor in the primed, comoving coordinates.

(f) In the context of an expanding Universe, give a brief physical interpretation of the scaling of the non-zero components of $T^{\alpha'\beta'}$ with the scale factor $a(t')$.

4

Geodesics

4.1 The Action and Lagrangian for a Curve

We have been talking sensibly, but vaguely, about straight lines and their close cousins geodesics in various spaces and space-time since the beginning of the book. It is now time to bolt this structure to firmer foundations. In this section and Section 4.2, we will develop a mathematical recipe for finding the geodesics of a space.

In space, a geodesic is the straightest possible curve between two fixed points. For flat spaces, this lines up with our intuition for a straight line. In a curved space, such as the surface of an apple, we can find geodesics by taking a piece of elastic and stretching it between the two points between which we wish to find the geodesic. The elastic band prefers its ground state, in which it stores the minimum amount of energy, and settles in the state where it has the minimum length, constrained by the curved surface. Of course, this does not work so well for some curved surfaces, a saddle for example, where the rubber will take a short cut between opposite ends of the saddle, leaving the surface above the valley, but we have to imagine that the elastic is constrained to lie in the surface, and then the method still works.

However, we could do with a more mathematically based way of finding the geodesics in spaces, and this is provided by the method of Lagrangian mechanics, though not the familiar rather recipe-book '$L = T - V$' version you were taught in classical mechanics. We will need to work out what the appropriate Lagrangian is in the case of general relativity and also how to get to the Euler–Lagrange equations from this Lagrangian. The solutions of these Euler–Lagrange equations will be the geodesics in the space, and once we generalise to space-time, the geodesics will be the paths taken by objects in free fall in a gravitational field.

Let us begin by considering two points A and B in a two-dimensional flat space. This is illustrated in Figure 4.1

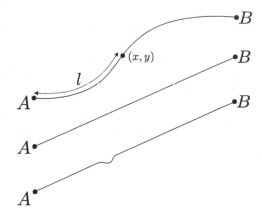

Figure 4.1 Three different pathways between two points A and B in a flat two-dimensional space.

The most general pathway is an arbitrary curve, illustrated by the uppermost path. I have superposed a point on the line, and give its Cartesian coordinates (x, y). The distance I have travelled along the line from point A, taken as the starting point, is l, which increases from $l = l_A$ at point A, reaching its maximum value of $l = l_B$ at point B. Below this general path, I have shown the intuitive solution for the geodesic, in flat space simply a straight line between A and B. This is the path of minimal length. The total length S of the curve between A and B can be written as the integral

$$S = \int_{l_A}^{l_B} dl, \tag{4.1}$$

where the integral is understood to be a path integral along the curve, which in general of course can be any curve. The straight line is the curve for which this integral is minimal. If you start with the straight line and then make any small variation from that solution, as illustrated in the lowest curve of Figure 4.1, then that increases the value of S by making the path slightly longer. So the straight line path is a local minimum of the functional S in the space of all possible paths between A and B. In some spaces, there is more than one local minimum of this integral and more than one geodesic between two points; on a sphere, for example, there are two geodesics connecting any two points on the spherical surface. However, in flat space, there is only one straight line between any two points.

The quantity S is an example of an action. In this case, it is simply the length of the curve, and it takes a different value for every curve and minimum value for the straight line.

We can write S in terms of Cartesian coordinates as follows:

$$S = \int_{l_A}^{l_B} dl$$

$$= \int_{l_A}^{l_B} \sqrt{dx^2 + dy^2}$$

$$= \int_{l_A}^{l_B} \sqrt{\left(\frac{dx}{dl}\right)^2 dl^2 + \left(\frac{dy}{dl}\right)^2 dl^2}$$

$$= \int_{l_A}^{l_B} \sqrt{\left(\frac{dx}{dl}\right)^2 + \left(\frac{dy}{dl}\right)^2} \, dl. \tag{4.2}$$

We now define the Lagrangian as the integrand of this path integral along the curve:

$$L = \sqrt{\left(\frac{dx}{dl}\right)^2 + \left(\frac{dy}{dl}\right)^2}. \tag{4.3}$$

This Lagrangian is a function of dx/dl and dy/dl. In curvilinear coordinate systems, there will be occurrences of the equivalents of the coordinates themselves, as well as their derivatives with respect to l. In Problem 4.1, for example, we will show that the Lagrangian in polar coordinates is

$$L = \sqrt{\left(\frac{dr}{dl}\right)^2 + r^2 \left(\frac{d\theta}{dl}\right)^2}. \tag{4.4}$$

4.2 The Euler–Lagrange Equations

Because we are interested in finding geodesics in general curved spaces in curvilinear coordinates, we leave the specific case of Cartesian coordinates for a while and analyse the general problem of minimising the length of the curve S for any coordinate system covering the space. This coordinate system has two coordinates (u, v), and the length of the curve is given by

$$S = \int_{l_A}^{l_B} L\left(u, \frac{du}{dl}, v, \frac{dv}{dl}\right) dl, \tag{4.5}$$

where again l_A and l_B are the values of the parameter l at the two end points A and B.

Referring again to Figure 4.1, if we start with an arbitrary curve and consider small excursions about that curve, then for at least some excursions, making the same sized small excursion on either side of the curve will result in a lengthening of the curve for the excursion on one side and a shortening of the curve for the same excursion on the opposite side. If we use a small parameter, say ε, to denote the amplitude of the excursion, where positive and negative ε denote excursions to the two different sides, then the change in the action, or curve length S, is linear in ε; it gets larger for one sign of ε and smaller for the other sign.

Consider now the special case of the straight line path. Any excursion whatsoever makes S larger. Either sign of ε results in a lengthening of the curve and an increase in S. This means that S is proportional to the square of ε, or perhaps a higher power, but there is no excursion for which the change in S is linear in ε. If we are only considering one particular choice of excursion with amplitude ε, then we can write

$$\frac{dS}{d\varepsilon} = \lim_{\varepsilon \to 0} \frac{S(\varepsilon) - S(0)}{\varepsilon} = 0, \qquad (4.6)$$

because the numerator is proportional to ε^2, so that the fraction becomes ε and hence tends to zero in the limit. We might therefore write, using the usual notation for small changes, that for small ε,

$$\delta S = \frac{dS}{d\varepsilon} \delta\varepsilon = 0. \qquad (4.7)$$

Again, for the special case of a straight line, this analysis will be true for any excursion – a bump or pimple at any point on the line will make it longer, and for that matter so will any combination of arbitrary bumps and pimples. We therefore write that for the straight line path,

$$\delta S = \delta \int_{l_A}^{l_B} L\left(u, \frac{du}{dl}, v, \frac{dv}{dl}\right) dl = 0. \qquad (4.8)$$

We are going to get tired of writing all those u, v, and their derivatives, so again we adopt an index notation, where for this two-dimensional example, i is either 1 for u or 2 for v:

$$\delta S = \int_{l_A}^{l_B} L\left(u^i, \frac{du^i}{dl}\right) dl = 0. \qquad (4.9)$$

The next step is to realise that any change in S that may happen as a result of altering the path is cumulative; to get the overall change in S, you sum up the changes resulting in whatever has been done to distort the curve at each line segment along its length. It is useful to think of the line as made up of a fixed number of line segments each of which either stays the same length or gets a little bit longer when you stretch the line. In either case the change δL in length of any line segment is zero when the initial path is a straight line, by the same argument just applied to δS. We therefore write, something subtly different in terms of symbols, which however leads to a way of progressing the mathematics substantially,

$$\delta S = \int_{l_A}^{l_B} \delta L\left(u^i, \frac{du^i}{dl}\right) dl = 0. \tag{4.10}$$

To understand the next step, it is important to be clear as to why L is a function of u^i and du^i/dl separately. Think of making two pen marks on a rubber band and then stretching it; then the pen marks get further apart. You could of course specify the distance between pen marks by taking the difference between their coordinates, so, for example, $u^{i+1} - u^i$. However, if there are a lot of very small sections, then these differences are handled in calculus using derivatives instead:

$$\frac{du^i}{dl} = \lim_{\delta l \to 0} \frac{u^{i+1} - u^i}{\delta l}$$

$$\frac{du^i}{dl}\delta l \simeq u^{i+1} - u^i, \tag{4.11}$$

where on the second line the equality is exact again as $\delta l \to 0$. You can see from this equation that the difference between the coordinates of adjacent pen marks on our rubber band is equivalent to the derivative of the coordinates at one end of the segment. So, rather than getting the total length by adding up the displacements between all the pen marks, we specify the coordinate u^i and its rate of change with respect to l at a set of points along the line, in the limit of an infinite number of points, and then do an integral to calculate the total length.

This method has the advantage that the Lagrangian is a local function of position, since u^i and du^i/dl are evaluated at the same point. It does however mean that we must treat u^i and du^i/dl as separate variables that can both change independently of each other. In fact, for the particular case of a line in flat space, only the derivatives du^i/dl appear in the Lagrangian. It is in non-Cartesian coordinates that factors of u^i also start to appear, and in these more general coordinate systems, the total change δL in the length of a given line segment can in general arise from changes in u^i and du^i/dl. We use the total derivative to find an expression for δL in terms of variations in these distinct variables:

$$\delta L = \left(\frac{\partial L}{\partial u^i}\right)\delta u^i + \frac{\partial L}{\partial (du^i/dl)}\delta\left(\frac{du^i}{dl}\right). \tag{4.12}$$

Now we reverse the variation and the derivative with respect to l in the second term:

$$\delta L = \left(\frac{\partial L}{\partial u^i}\right)\delta u^i + \frac{\partial L}{\partial (du^i/dl)}\frac{d}{dl}\delta u^i. \tag{4.13}$$

Next, we apply this variation to the action integral and integrate the more complicated second term by parts. Although you will have integrated by parts many times, this application requires a thorough understanding of how it works. In fact, it is just an application of the product rule of differentiation, applied in reverse! Consider the derivative with respect to l of a slightly different expression:

$$\frac{d}{dl}\left(\frac{\partial L}{\partial (du^i/dl)}\,\delta u^i\right) = \delta u^i\,\frac{d}{dl}\left(\frac{\partial L}{\partial (du^i/dl)}\right) + \frac{\partial L}{\partial (du^i/dl)}\,\frac{d}{dl}\,\delta u^i. \qquad (4.14)$$

This is different from the expression in Equation (4.13) as the operator d/dl is applied to both the partial derivative and the small change δu^i. We can rearrange this expression by moving the first term on the right to the left-hand side:

$$\frac{d}{dl}\left(\frac{\partial L}{\partial (du^i/dl)}\,\delta u^i\right) - \delta u^i\,\frac{d}{dl}\left(\frac{\partial L}{\partial (du^i/dl)}\right) = \frac{\partial L}{\partial (du^i/dl)}\,\frac{d}{dl}\,\delta u^i. \qquad (4.15)$$

The expression on the right is the second term in the total derivative from Equation (4.12), so we substitute back in for it:

$$\delta L = \left(\frac{\partial L}{\partial u^i}\right)\delta u^i + \frac{d}{dl}\left(\frac{\partial L}{\partial (du^i/dl)}\,\delta u^i\right) - \delta u^i\,\frac{d}{dl}\left(\frac{\partial L}{\partial (du^i/dl)}\right). \qquad (4.16)$$

Now this variation in L must be substituted back into the action integral of Equation (4.8). Consider in particular this integral for the middle term on the right:

$$\int_{l_A}^{l_B}\frac{d}{dl}\left(\frac{\partial L}{\partial (du^i/dl)}\,\delta u^i\right)dl. \qquad (4.17)$$

Any time you have the integral with respect to some variable of the derivative of some quantity with respect to that same variable, the result is the quantity evaluated at the end points of the integral. This is sometimes called the fundamental theorem of calculus, and it is closely related to the definition of the derivative of a function, but again thinking about the concept in reverse! Therefore you can just replace this integral as follows:

$$\int_{l_A}^{l_B}\frac{d}{dl}\left(\frac{\partial L}{\partial (du^i/dl)}\,\delta u^i\right)dl = \left[\frac{\partial L}{\partial (du^i/dl)}\,\delta u^i\right]_{l_A}^{l_B}. \qquad (4.18)$$

However, the displacements δu^i at the ends of the path are zero; we are only considering the deformations of the path between these points, not at them. Therefore this term is zero. This zero term is called a surface term, a reference to the same phenomenon in higher dimensions where a volume integral is bounded by a surface in the same way that a line is bounded by two points. We are left with only the two outer terms from Equation (4.16) contributing to the δS:

$$\delta S = \int_{l_A}^{l_B} \left(\frac{\partial L}{\partial u^i} \right) \delta u^i - \delta u^i \frac{d}{dl} \left(\frac{\partial L}{\partial (du^i/dl)} \right) dl = 0$$

$$= \int_{l_A}^{l_B} \left[\frac{\partial L}{\partial u^i} - \frac{d}{dl} \left(\frac{\partial L}{\partial (du^i/dl)} \right) \right] \delta u^i \, dl = 0. \tag{4.19}$$

For this expression to be zero, the integrand must be zero, and hence we arrive at the following conditions for the path joining A and B to be a straight line:

$$\frac{\partial L}{\partial u^i} - \frac{d}{dl} \left(\frac{\partial L}{\partial (du^i/dl)} \right) = 0. \tag{4.20}$$

These are the Euler–Lagrange equations. You have met them before in classical mechanics, but perhaps you have never seen them derived. Also, note that the quantity being minimised here, the distance between two points, is not the same quantity that is minimised when using the Lagrangian method to solve problems in classical mechanics. There we were minimising $T - V$, the difference between the kinetic and potential energies of a particle. We will see how this approach is related to our Lagrangian for the length of a curve in Problem 4.4.

4.3 Geodesics in Flat Space

We now return to our specific problem, showing that a straight line is the shortest distance between two points in a flat two-dimensional space using Cartesian coordinates. The Lagrangian is that of Equation (4.3),

$$L = \sqrt{\left(\frac{dx}{dl} \right)^2 + \left(\frac{dy}{dl} \right)^2}. \tag{4.21}$$

The partial deriatives $\partial L/\partial x$ and $\partial L/\partial y$ are both zero, not forgetting that dx/dl and dy/dl are considered independent variables from x and y. The partial derivatives with respect to dx/dl and dy/dl are

$$\frac{\partial L}{\partial (dx/dl)} = \frac{1}{2} \left(\left(\frac{dx}{dl} \right)^2 + \left(\frac{dy}{dl} \right)^2 \right)^{-1/2} \times 2\frac{dx}{dl}$$

$$= \frac{1}{L} \frac{dx}{dl}$$

$$\frac{\partial L}{\partial (dy/dl)} = \frac{1}{L} \frac{dy}{dl}. \tag{4.22}$$

Now we are going to have to compute d/dl of each of these partial derivatives. This looks at first sight to be a complex job, because L is a function of both dx/dl and

dy/dl. However, there is a great simplification that can be arrived at by examining the derivative of L with respect to l in a bit more detail.

Recalling that L is a function of dx/dl and dy/dl, we can write its total derivative as

$$\delta L = \frac{\partial L}{\partial(dx/dl)}\,\delta\frac{dx}{dl} + \frac{\partial L}{\partial(dy/dl)}\,\delta\frac{dy}{dl}. \tag{4.23}$$

Therefore the ordinary derivative with respect to l can be written

$$\frac{dL}{dl} = \frac{\partial L}{\partial(dx/dl)}\frac{d(dx/dl)}{dl} + \frac{\partial L}{\partial(dy/dl)}\frac{d(dy/dl)}{dl}$$
$$= \frac{1}{L}\left[\frac{dx}{dl}\frac{d^2x}{dl^2} + \frac{dy}{dl}\frac{d^2y}{dl^2}\right]. \tag{4.24}$$

Now let us consider a short segment of our curve, as illustrated in Figure 4.2. In terms of the angle θ the various derivatives in Equation (4.24) are

$$\frac{dx}{dl} = \cos\theta \qquad\qquad \frac{dy}{dl} = \sin\theta$$
$$\frac{d^2x}{dl^2} = -\sin\theta\frac{d\theta}{dl} \qquad\qquad \frac{d^2y}{dl^2} = \cos\theta\frac{d\theta}{dl}. \tag{4.25}$$

Substituting these results into Equation (4.24), we get

$$\frac{dx}{dl}\frac{d^2x}{dl^2} + \frac{dy}{dl}\frac{d^2y}{dl^2} = -\cos\theta\sin\theta\frac{d\theta}{dl} + \sin\theta\cos\theta\frac{d\theta}{dl} = 0. \tag{4.26}$$

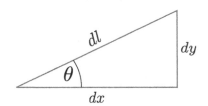

Figure 4.2 A short segment of a curve in a flat two-dimensional space.

Therefore we conclude that

$$\frac{dL}{dl} = 0. \tag{4.27}$$

This result, derived here for the special case of Cartesian coordinates in a flat space, turns out to be quite general and significant. The generality comes from the fact that both L and l are tensors of rank $^0_0$. For example, they do not change when you change to a new coordinate system rotated with respect to the original one.

Everything in Equation (4.27) transforms as the components of a tensor, and, consequently, dL/dl is zero in any other coordinate system, not just Cartesian coordinates. This type of argument will occur many times in the rest of the book and is a central reason why tensor methods are so useful. You can come up with a result in a coordinate system where you have intuition, write that result in terms of tensor components, and the result generalises to arbitrary coordinates where your intuition may let you down.

It is also useful to look at this result a different way. We substitute $dx/dl = \cos\theta$ and $dy/dl = \sin\theta$ into the Lagrangian and obtain

$$
\begin{aligned}
L &= \sqrt{\left(\frac{dx}{dl}\right)^2 + \left(\frac{dy}{dl}\right)^2} \\
&= \sqrt{\cos^2\theta + \sin^2\theta} \\
&= 1.
\end{aligned}
\tag{4.28}
$$

We will come back to $L = 1$ when discussing cyclic coordinates in Section 4.6. It may seem strange that the quantity L that we are making all this fuss over turns out to be 1. In fact, this is due to our very specific, and special, choice of the parameter. We are just minimising the length of the curve by adding up the lengths of lots of small pieces. The result of that sum depends on the number of pieces if each piece is the same length as all the others. In general, the parameter that describes how far along the curve you have travelled does not have to be l. All that is required is that this parameter is 'monotonically increasing', meaning that its value is always getting larger as you move along the curve in one direction. For our particular choice of the parameter l, you can imagine an ordinary flexible tape measure laid along the curve with its graduations regularly spaced. The regularly spaced graduations mean that the definition of the change in l, $\delta l = \delta L$, is the same for all values of l, and hence for this particular choice of the parameter, $dL/dl = 0$.

We could if we wanted to choose some other parameter, where the definition of δL for the same sized step is in fact different at different values of l. Again, imagine a second tape measure marked off in some peculiar units where the gap between gradations changes as you move along this rather strange tape measure. It is still a perfectly good monotonically increasing measure of where you are, but in terms of this contrived parameter, call it l', we find that $dL/dl' \neq 0$. This is inevitably true if the chosen parameter is not invariant with respect to coordinate transformations. We will encounter a class of problems where we are forced to employ such a parameter in Section 4.5.

For now, sticking with l as the distance along the curve, we can treat L as a constant in differentiation with respect to L and write the Euler–Lagrange equations in x and y as

$$\frac{d}{dl}\left(\frac{\partial L}{\partial(dx/dl)}\right) = \frac{d}{dl}\frac{1}{L}\frac{dx}{dl}$$

$$= \frac{1}{L}\frac{d^2x}{dl^2} = 0$$

$$\frac{d}{dl}\left(\frac{\partial L}{\partial(dy/dl)}\right) = \frac{d}{dl}\frac{1}{L}\frac{dy}{dl}$$

$$= \frac{1}{L}\frac{d^2y}{dl^2} = 0. \tag{4.29}$$

Since L is not zero, we can divide it out of both equations and obtain

$$\frac{d^2x}{dl^2} = 0 \qquad\qquad \frac{d^2y}{dl^2} = 0. \tag{4.30}$$

We solve these equations by first integrating each of them once to get

$$\frac{dx}{dl} = P \qquad\qquad \frac{dy}{dl} = Q, \tag{4.31}$$

where P and Q are constants. Then we can eliminate the parameter l using the chain rule as follows:

$$\frac{dy}{dx} = \frac{dy}{dl}\frac{dl}{dx}$$

$$= \frac{Q}{P} = M, \tag{4.32}$$

where we have named the ratio of Q to P as another constant M. Integrating the expression for dy/dx once, we obtain

$$y = Mx + C, \tag{4.33}$$

where C is a further constant. This is the equation of a straight line, so we have shown that the Lagrangian formalism correctly identifies straight lines as the shortest distances between points in a two-dimensional flat space described by Cartesian coordinates.

4.4 Tensor Expression for the Lagrangian

The Lagrangian we used in Cartesian coordinates in Sections 4.1 and 4.3 can be written in index notation as

$$L = \sqrt{\eta_{ij}\frac{dx^i}{dl}\frac{dx^j}{dl}}, \tag{4.34}$$

where η_{ij} is 1 when $i = j$ and 0 when $i \neq j$, and, as before, $x^1 = x$ and $x^2 = y$. There are implied sums over i and j each taking the values 1 or 2 because of the

Einstein summation convention. We may turn this expression into a tensor one by noting that the η_{ij} are special cases of the metric coefficients for Cartesian coordinates. Therefore the Lagrangian can also be written

$$L = \sqrt{g_{ij}\frac{dx^i}{dl}\frac{dx^j}{dl}}. \tag{4.35}$$

In this form, the Lagrangian is a tensor. Because all its indices are contracted, it is of rank $^0_0$, so it is also an invariant. We can use it to find the geodesics for a flat space in curvilinear coordinates, such as plane polar or plane hyperbolic coordinates, or to find the geodesics in curved spaces where we will no longer be able to use ordinary Cartesian coordinates to cover the space.

One more thing, or result from Equation (4.27). The small displacement dl between nearby points in the plane is an invariant because it is just a displacement and remains the same no matter which coordinate system you are using to write its coordinates. Therefore $dL/dl = 0$ is a tensor expression and is true in any coordinate system. This means that whenever we are faced with an expression containing an L that we are taking the ordinary derivative of with respect to an invariant parameter, such as the distance down the curve, l, then we may treat the L as a constant with respect to the differentiation just as we did in Section 4.3. The resulting simplification makes it easier to derive the Euler–Lagrange equations. We will soon see that this simplification can, with an appropriate choice of parameter, also be used in relativistic problems in space-time.

4.5 Lagrangians in Space-Time

The modifications to the discussion in Sections 4.1 and 4.3 on straight lines in a flat space carry over in Section 4.4 to the Lagrangian for finding the geodesics in curved spaces. The transition to space-time is actually not terribly dramatic either. The main modification concerns what the curves physically mean. They are now the world lines of particles, and they exist in four-dimensional space-time. This means that we are going to have to think about a suitable parameter to describe positions particles along world lines.

Consider first a parameter to describe the trajectories of freely falling particles. If we write down the interval between two events on the world line of the particle, is it given in general by

$$ds^2 = -c^2\,dt^2 + dx^2 + dy^2 + dz^2. \tag{4.36}$$

Because our observer is moving with the particle as it travels, the origin of their coordinate system always has the particle at it, so that $dx^1 = dx^2 = dx^3 = 0$ and

$dt = d\tau$. Therefore we have

$$ds^2 = -c^2 \, d\tau^2. \tag{4.37}$$

The only non-zero component is the time component $d\tau$ as it falls freely. You can imagine that the observer falling with the particle is wearing a watch; the time on the dial of this watch is the proper time τ in the frame of the particle. It is this time τ that is used as a parameter in general relativity. It is a Lorentz invariant, a tensor of rank $^0_0$, because you can always make a coordinate transformation into a frame falling with the particle and look at the time passing on your wristwatch. This is the proper time coordinate used for the particle.

A second adjustment comes about because of the definition of the $\eta_{\mu\nu}$, the metric in a freely falling frame, so that it has a minus sign on the 00 component. Because of this minus sign, the interval ds^2 between two neighbouring events on the world line of a particle is always negative for particles travelling at less than c (which is all of them.). Consequently, so that the square root in the Lagrangian is of a positive number, we need to embed a compensating minus sign under the square root in the Lagrangian for world lines of freely falling particles. This leads us to the expression for the space-time Lagrangian,

$$L = \sqrt{-g_{\mu\nu} \frac{dx^\mu}{d\tau} \frac{dx^\nu}{d\tau}}. \tag{4.38}$$

By a method analogous to that given in Section 4.3 we can show (see Problem 4.2) that

$$\frac{dL}{d\tau} = 0, \tag{4.39}$$

so that we can continue to treat L like a constant when we differentiate expressions containing it with respect to τ, just as we did when the parameter was l in space rather than space-time.

There are classes of observers for which there are caveats and exceptions. The first class where you have to be careful are freely falling photons and other massless particles. Such particles move at the speed of light, and consequently all events in the rest frame of such particles happen simultaneously. For particles in these frames moving at speed c, known as Breit frames, the proper time stands still, and it is not practical to use τ as the parameter to describe position on a world line.

For the world lines of photons, you instead choose some other parameter to describe where you are on the world line of the particle as it falls. All that is really required is that the parameter is monotonically increasing (getting larger) as time moves forward. The time in the rest frame of another observer not moving at the speed of light is an acceptable choice. Give this parameter the symbol t_o for 'observer time', and we can then write our Lagrangian as

$$L = \sqrt{-g_{\mu\nu} \frac{dx^\mu}{dt_o} \frac{dx^\nu}{dt_o}}. \qquad (4.40)$$

Sometimes the parameter t_o is called an affine parameter. This name comes about because the position of the zero of t_o does not matter; indeed, very little about this parameter matters other than the requirement that it increases along the trajectory. However, in many ways, affine parameters are not as good as proper times. For example, affine time intervals dt_o are not tensors, and consequently dL/dt_o is not zero. This means that you cannot treat L as a constant when differentiating with respect to t_o.

There are compensations for photons. You can exploit the fact that proper time intervals in the rest frame of particles moving at the speed of light are zero, so that you can write

$$ds^2 = g_{\mu\nu} \, dx^\mu \, dx^\nu = 0. \qquad (4.41)$$

This is true in the frame of reference of any observer, freely falling or not. This fact often provides an alternative way to deduce the geodesics of photons, as these equations connecting time and spatial intervals provide a pathway to solving for the particle trajectory that is not available for massive particles moving at less than the speed of light.

The other class of particles for which things are more complicated are particles that are not on freely falling trajectories. For these particles, the proper time interval between events is not an invariant because of the metric coefficients. We have

$$ds^2 = g_{00} \, c^2 \, dt_a^2, \qquad (4.42)$$

where t_a is the time between events to an observer accelerating with the moving particle. So, although you can still use the rest frame time t_a as a parameter, it is not an invariant under changes of observer. This means, again, that dL/dt_a is not zero, and there is no getting around the complexity of the Euler–Lagrange equations for non-freely falling particles in the presence of gravitational fields. However, in this course, we will only be concerned with calculating the trajectories of particles and fields that are in exact or approximate free fall. It is the observers from whose perspectives the particle behaviours are measured who will often be accelerating, but the Euler–Lagrange equations, written with the proper time as a parameter, will still yield the correct observed motion for the particle in the coordinate system of the observer, including how the observer time is related to proper time.

4.6 Cyclic Coordinates

Confining ourselves to particles in free fall, so that the proper time is a Lorentz invariant, we now discuss cyclic coordinates. We start by writing down again the general form of the Euler–Lagrange equations, this time in space-time:

$$\frac{\partial L}{\partial x^\mu} - \frac{d}{d\tau}\frac{\partial L}{\partial(dx^\mu/d\tau)} = 0. \tag{4.43}$$

We now write down again the form of the Lagrangian

$$L = \sqrt{-g_{\mu\nu}\frac{dx^\mu}{d\tau}\frac{dx^\nu}{d\tau}}. \tag{4.44}$$

Notice that the only way for dL/dx^μ to be zero is for x^μ not to appear explicitly in the coefficients of the metric. Coordinates that do not appear explicitly in the metric are called cyclic.

For example, going back to the two-dimensional space, the metric coefficients on a flat plane are $g_{xx} = 1$, $g_{xy} = g_{yx} = 0$, and $g_{yy} = 1$. The Lagrangian corresponding to this metric is

$$L = \sqrt{\left(\frac{dx}{dl}\right)^2 + \left(\frac{dy}{dl}\right)^2}. \tag{4.45}$$

This Lagrangian contains no explicit dependence on x or y, and therefore you can write that two quantities are constant:

$$\frac{\partial L}{\partial(dx/d\tau)} \quad \text{and} \quad \frac{\partial L}{\partial(dy/d\tau)}. \tag{4.46}$$

What are these conserved quantities? They are

$$\frac{\partial L}{\partial(dx/dl)} = \frac{dx/dl}{L} = \frac{dx}{dl}$$
$$\frac{\partial L}{\partial(dy/dl)} = \frac{dy/dl}{L} = \frac{dy}{dl}, \tag{4.47}$$

where in the last step, we have used the fact that $L = 1$ when the parameter l is an invariant length. The fact that dx/dl and dy/dl are constants, conserved quantities, along the trajectory is expected if the trajectory is a straight line. Conserved quantities can be used as an alternate route to deducing the geodesics without solving the full Euler–Lagrange equations in the case where you have cyclic coordinates.

One word of warning. The conserved quantities that you find are dependent on what coordinate system you choose. If we had started in plane polar coordinates, then we would have discovered another conserved quantity that is not obvious from the same physics in Cartesian coordinates. However, we might have missed the two conserved quantities from Equation (4.47) that follow from the cyclic coordinates present in Cartesian coordinates. A more methodical way of uncovering conserved quantities is to use the technique of Killing vectors, but as this book aims to uncover some of the material rather than covering all of it, this is a topic that I have elected to omit.

Let us generalise this statement about the link between cyclic coordinates and conserved quantities. Suppose in space-time that a cyclic coordinate is identified and it is x^μ for one particular value of μ. This means that the coordinate x^μ does not appear explicitly in the metric coefficients. This means that the quantity

$$v_\mu = \frac{\partial L}{\partial(dx^\mu/d\tau)} \tag{4.48}$$

is conserved. Notice that I have written the conserved quantity with a lower index. This is because the index quantity on the right-hand side appears in the denominator, and an upper index on a tensor in the denominator will transform as a tensor of rank $^0_1$. We can however always find expressions for the components of v_μ even if they are not all conserved quantities. We can then use

$$v^\mu = g^{\mu\nu} v_\nu \tag{4.49}$$

with an implied sum over ν to determine the components of v^μ. In many cases, notably those of the Schwarzschild geometry and the Friedmann–Robertson–Walker metric studied in Chapters 6 and 7, the metric is diagonal, $g_{\mu\neq\nu} = 0$, so that the relationships between v^μ and v_μ involve a simple multiplicative factor rather than a sum over components. We can then exploit the fact that $v_\mu v^\mu$ is an invariant to solve for the motion. This technique is applied in Problem 4.3 to prove the energy–momentum–mass relation for a freely propagating particle in the absence of a gravitational field and in Problem 6.7 to calculate the precession of the perihelion of Mercury.

4.7 Differentiation of Tensors

We are going to work with derivatives of the components of tensors. We have already encountered the first of these, the derivative of a tensor of rank $^0_0$, which we discussed in Section 2.8. We found that the components of the four spatial derivatives of a scalar field $\Phi(x)$ transform as a tensor of rank $^0_1$:

$$\partial_{\beta'}\Phi = \frac{\partial x^\alpha}{\partial x^{\beta'}}\partial_\alpha\Phi. \tag{4.50}$$

What about the derivatives of the components of tensors of higher rank? Are these the components of tensors? It turns out that they are not. This is because a derivative is the comparison of a function of position at two neighbouring points. When that function of position is, say, a vector, then between two neighbouring points, two things can change. First, the components of the vector in some coordinate system can change. Second, the basis vectors along which those components run can change too. It turns out that the vector as a whole, like the scalar field, does have a tensor derivative. However, the components of the vector have derivatives that do not transform like tensors. The components are only part of the story, and hence they cannot transform tensorially like the whole object.

Let us analyse one example, that of a vector. Suppose we have a vector field $\vec{A}(x)$. This could be a vector field either in a flat space or in a curved one, and also in any number of spatial or space-time dimensions. It is a vector that takes a different value, in both magnitude and direction, at every point in space or space-time. We can differentiate this vector field with respect to any of the coordinates, resulting in the partial derivatives

$$\frac{\partial \vec{A}}{\partial x^\alpha}. \tag{4.51}$$

We can also use the chain rule to connect partial spatial derivatives to those observed in a different frame of reference,

$$\frac{\partial \vec{A}}{\partial x^{\beta'}} = \frac{\partial x^\alpha}{\partial x^{\beta'}} \frac{\partial \vec{A}}{\partial x^\alpha}. \tag{4.52}$$

This is exactly the same analysis we did for a scalar field Φ. It shows that the spatial partial derivatives of the vector field $\vec{A}$ transform as tensors of rank $\binom{0}{1}$, just as the partial derivatives of the scalar field do. However, there is a complication with vector fields when we consider the vector not as an arrow in space, but as a quantity expressed in terms of its components in some coordinate system. We write

$$\vec{A} = A^\mu \vec{e}_\mu. \tag{4.53}$$

Taking the same spatial derivative, we now end up using the product rule:

$$\begin{aligned}
\frac{\partial \vec{A}}{\partial x^\alpha} &= \frac{\partial}{\partial x^\alpha} A^\mu \vec{e}_\mu \\
&= \vec{e}_\mu \frac{\partial A^\mu}{\partial x^\alpha} + A^\mu \frac{\partial \vec{e}_\mu}{\partial x^\alpha}.
\end{aligned} \tag{4.54}$$

The tensor transformation properties of $\vec{A}$ are the same as they were before – the quantity on the left of the equals sign transforms as a tensor of rank $\binom{0}{1}$. However, there are two objects on the right that we have not yet considered. The first is the spatial derivative of the component of a vector, $\partial A^\mu / \partial x^\alpha$. The second is the spatial derivative of the basis vector.

4.8 Christoffel Symbols

4.8.1 Christoffel Symbols in Curvilinear Coordinates

Let us consider what the spatial derivative of a basis vector might mean in flat space. Refer to Figure 4.3.

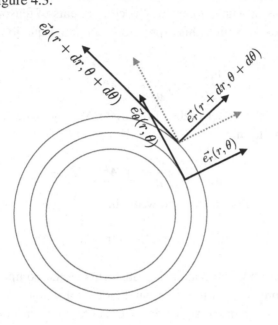

Figure 4.3 Basis vectors in polar coordinates at two nearby points.

Here are drawn representations of the basis vectors in the directions of increasing r and θ at two nearby points in a flat plane. Notice that the basis vectors at the two points are different. However, you can translate the basis vectors at point (r, θ) to the point $(r + dr, \theta + d\theta)$. Therefore you can express the change in each of the basis vectors between the two nearby points as linear coefficients of the basis vectors at the start point:

$$\frac{\partial \vec{e}_r}{\partial r} = \Gamma^r_{rr} \vec{e}_r(r, \theta) + \Gamma^\theta_{rr} \vec{e}_\theta(r, \theta)$$

$$\frac{\partial \vec{e}_r}{\partial \theta} = \Gamma^r_{r\theta} \vec{e}_r(r, \theta) + \Gamma^\theta_{r\theta} \vec{e}_\theta(r, \theta)$$

$$\frac{\partial \vec{e}_\theta}{\partial r} = \Gamma^r_{\theta r} \vec{e}_r(r, \theta) + \Gamma^\theta_{\theta r} \vec{e}_\theta(r, \theta)$$

$$\frac{\partial \vec{e}_\theta}{\partial \theta} = \Gamma^r_{\theta\theta} \vec{e}_r(r, \theta) + \Gamma^\theta_{\theta\theta} \vec{e}_\theta(r, \theta). \tag{4.55}$$

The coefficients Γ^k_{ij} are called Christoffel symbols, or sometimes connection coefficients, or collectively just the affine connection. Generalising to arbitrary space or space-time dimensions, the definition in index notation is

$$\frac{\partial \vec{e}_\mu}{\partial x^\nu} = \Gamma^\alpha_{\mu\nu} \vec{e}_\alpha.$$

(4.56)

Here there is an implied sum over α. We do not yet know how to calculate the Christoffel symbols or whether they are the components of tensors, though in fact they are not. We next substitute this expression back into Equation (4.54) and arrive at

$$\frac{\partial \vec{A}}{\partial x^\alpha} = \vec{e}_\mu \frac{\partial A^\mu}{\partial x^\alpha} + A^\mu \Gamma^\beta_{\mu\alpha} \vec{e}_\beta.$$

(4.57)

Now we use one of the tensor tricks of the trade. We swap the repeated indices μ and β in the second term:

$$\frac{\partial \vec{A}}{\partial x^\alpha} = \vec{e}_\mu \frac{\partial A^\mu}{\partial x^\alpha} + A^\beta \Gamma^\mu_{\beta\alpha} \vec{e}_\mu.$$

(4.58)

This allows us to factor out the $\vec{e}_\mu$ and write this as

$$\frac{\partial \vec{A}}{\partial x^\alpha} = \vec{e}_\mu \left(\frac{\partial A^\mu}{\partial x^\alpha} + \Gamma^\mu_{\beta\alpha} A^\beta \right).$$

(4.59)

As discussed before, the left-hand side transforms as the components of a tensor of rank $^0_1$. Therefore so does the right-hand side as a whole. Written like this, it looks like something, in brackets, that is contracted with a tensor of rank $^0_1$. For the right-hand side as a whole to be a tensor of rank $^0_1$, the quantity in brackets which it is contracted with must transform as a tensor of rank $^1_1$. Yet, the quantity in brackets contains in part the partial derivative of the components A^μ of a vector field with respect to displacements x^α. This is a modified derivative of these components that does have a tensor transformation law. We call it the Lorentz covariant derivative of a rank $^1_0$ tensor, or just the covariant derivative. It has the symbol

$$D_\alpha A^\mu = \frac{\partial A^\mu}{\partial x^\alpha} + \Gamma^\mu_{\beta\alpha} A^\beta.$$

(4.60)

It turns out that Christoffel symbol terms can be added to, or subtracted from, the spatial derivatives of the components of tensors of arbitrary rank to form derivatives that transform as tensors. The recipe is different for each tensor rank, but the recipes follow a pattern that will be come clear when we find the covariant derivatives of components of a few more tensor types.

Another property of Christoffel symbols is that they are symmetric with respect to interchange of their two lower indices. To prove this, we can write a basis vector as

$$\vec{e}_\beta = \frac{\partial \vec{r}}{\partial x^\beta}.$$

(4.61)

We differentiate this expression with respect to x^γ and substitute in for the derivative of the basis vector from Equation (4.56) with the Christoffel symbols:

$$\frac{\partial \vec{e}_\beta}{\partial x^\gamma} = \Gamma^\delta_{\beta\gamma} \vec{e}_\delta = \frac{\partial^2 \vec{r}}{\partial x^\gamma \partial x^\beta}. \tag{4.62}$$

The right-hand side is symmetric with respect to the exchange $\beta \leftrightarrow \gamma$ by the equality of mixed partial derivatives. Therefore the Christoffel symbol is also symmetric with respect to the same interchange. This property will be useful presently in finding a formula for the Christoffel symbols in terms of the metric coefficients.

4.8.2 Christoffel Symbols in Curved Spaces

The above argument was made based on thinking about curvilinear coordinates in a flat two-dimensional space. We are also going to want to differentiate in a curved space. How is that going to be possible? It will only work if we can think of the change in the basis vectors between two neighbouring points as a superposition of coefficients (the Christoffel symbols) times the basis vectors at the point where you are starting. This does not at first sight seem possible. Staying in two dimensions, we imagine a curved two-dimensional surface.

We learned in Chapter 1 about tangent spaces and about how at any point on a curved surface you can draw a tangent plane that touches the curved surface at that point. If you have two neighbouring points, as shown in Figure 4.4, then the tangent planes at these two points will not be the same.

How then is it possible to express the change in the basis vectors as a superposition of the basis vectors at point A? It turns out that what gets you out of trouble is that the intersection of two planes is a straight line. You can imagine folding a piece of paper and positioning it so that two points on the paper touch the curved space in two different places. Now we can imagine marking the two points and the basis vectors at those points, then flattening the paper onto a table, and proceeding just as we did before in curvilinear coordinates in a flat space. From then on, the procedure is exactly the same. This trick of connecting the tangent spaces at neighbouring points makes differential geometry work. Effectively, you can show that two close-by points in a curved space share a tangent plane. We will encounter this again when we study curvature of spaces in the next chapter.

4.9 Covariant Derivatives

4.9.1 Review of Rank $^1_0$

In Section 4.8, we introduced the Christoffel symbols and learned that they can be used to make a covariant derivative of the components A^μ of a tensor of rank $^1_0$,

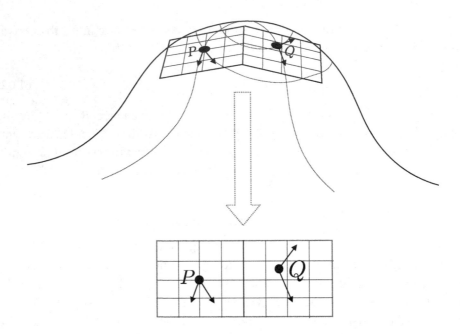

Figure 4.4 A folded piece of paper touching two close-by points on a curved surface. Superposed on the curved space, I have shown a contour of an arbitrary curvilinear coordinate system and the basis vectors in the direction of increasing values of the coordinates. Superposed on the folded paper is a flat space coordinate system with its Cartesian grid. You can see that the basis vectors at the two points lie both in the curved space and in the flat tangent space of the tangent paper. Note that here the chosen points are not that close-by; you have to imagine that P and Q are sufficiently close together that the basis vectors along the curvilinear coordinates are similar in magnitude and direction at the two points. If this is not true, then make P and Q closer together and repeat the argument.

whose components transform like those of a tensor of rank $\begin{smallmatrix}1\\1\end{smallmatrix}$,

$$D_\alpha A^\mu = \frac{\partial A^\mu}{\partial x^\alpha} + \Gamma^\mu_{\beta\alpha} A^\beta. \tag{4.63}$$

In this section, we work out how to write the covariant derivatives of the components of tensors of other ranks. As we will see, the definitions are all different, but they all involve the components of the tensors, their partial derivatives, and the Christoffel symbols, which we will see play a fundamental role in general relativity.

4.9.2 Rank $\begin{smallmatrix}0\\1\end{smallmatrix}$

We start out with the fact that scalars Φ have spatial derivatives $\partial_\mu \Phi$, which transform as the components of a tensor of rank $\binom{0}{1}$. However, the contraction of a tensor of rank $\binom{1}{0}$ with a tensor of rank $\binom{0}{1}$ is a scalar (a tensor of rank $\begin{smallmatrix}0\\0\end{smallmatrix}$), and hence if

$\Phi = A^\mu A_\mu$, then $\partial_\mu S = \partial_\mu A^\nu A_\nu$ is a tensor of rank $^0_1$. We can expand this tensor using the product rule, so that

$$\partial_\alpha (A^\mu A_\mu) = A_\mu \partial_\alpha A^\mu + A^\mu \partial_\alpha A_\mu. \tag{4.64}$$

We can write $\partial_\alpha A^\mu$ in terms of the covariant derivative, because $D_\alpha A^\mu = \partial_\alpha A^\mu + \Gamma^\mu_{\sigma\alpha} A^\sigma$. Substituting this into Equation (4.64), we obtain

$$\begin{aligned} \partial_\alpha (A^\mu A_\mu) &= A_\mu \left(D_\alpha A^\mu - \Gamma^\mu_{\sigma\alpha} A^\sigma \right) + A^\mu \partial_\alpha A_\mu \\ &= A_\mu D_\alpha A^\mu + A^\mu \partial_\alpha A_\mu - A_\mu \Gamma^\mu_{\sigma\alpha} A^\sigma. \end{aligned} \tag{4.65}$$

Swapping the σ and μ indices in the last term, we get

$$\begin{aligned} \partial_\alpha (A^\mu A_\mu) &= A_\mu D_\alpha A^\mu + A^\mu \partial_\alpha A_\mu - A_\sigma \Gamma^\sigma_{\mu\alpha} A^\mu \\ &= A_\mu D_\alpha A^\mu + A^\mu \left(\partial_\alpha A_\mu - \Gamma^\sigma_{\mu\alpha} A_\sigma \right). \end{aligned} \tag{4.66}$$

The brackets contain the partial derivative of A_μ with a correction term containing A_σ and a Christoffel symbol. If we call this the covariant derivative of the components A_μ of a tensor of rank $^0_1$,

$$D_\alpha A_\mu = \partial_\alpha A_\mu - \Gamma^\sigma_{\mu\alpha} A_\sigma, \tag{4.67}$$

then this looks right as an equivalent to Equation (4.63) for the components A^μ of a tensor of rank $^1_0$. The covariant derivative of the components of a tensor of rank $^0_1$ transforms like the components of a tensor of rank $^0_2$. As an added benefit, we can write a covariant derivative version of the product rule,

$$\partial_\alpha (A^\mu A_\mu) = A_\mu D_\alpha A^\mu + A^\mu D_\alpha A_\mu, \tag{4.68}$$

which looks like the ordinary product rule except that where the objects to be differentiated have partial derivatives that are not the components of tensors, these partial derivatives are replaced by covariant derivatives.

4.9.3 Rank $^0_2$

Let us do one more, because it turns out to be important. Consider a tensor of rank $\binom{0}{2}$, so with two lower indices. In particular, consider such a tensor having components consisting of the products of the components of two tensors each of rank $\binom{0}{1}$; this tensor is $R_{\nu\sigma} = A_\nu B_\sigma$. Now take a partial derivative of this tensor, $\partial_\lambda (A_\nu B_\sigma)$. This derivative is not a tensor, any more than $\partial_\lambda A_\nu$ is. But, we can expand this derivative using the product rule:

$$\begin{aligned} \partial_\lambda (A_\nu B_\sigma) &= A_\nu \partial_\lambda B_\sigma + B_\sigma \partial_\lambda A_\nu \\ &= A_\nu \left(D_\lambda B_\sigma + \Gamma^\omega_{\lambda\sigma} B_\omega \right) + B_\sigma \left(D_\lambda A_\nu + \Gamma^\omega_{\lambda\nu} A_\omega \right) \\ &= A_\nu D_\lambda B_\sigma + B_\sigma D_\lambda A_\nu + \Gamma^\omega_{\lambda\sigma} A_\nu B_\omega + \Gamma^\omega_{\lambda\nu} A_\omega B_\sigma. \end{aligned} \tag{4.69}$$

Now we assume that $D_\lambda(A_\nu B_\sigma)$ can be expanded with the chain rule, in which case we can re-write the last line as

$$\partial_\lambda(A_\nu B_\sigma) = D_\lambda(A_\nu B_\sigma) + \Gamma^\omega_{\lambda\sigma} A_\nu B_\omega + \Gamma^\omega_{\lambda\nu} A_\omega B_\sigma. \tag{4.70}$$

Re-writing in terms of $R_\nu\sigma$, we obtain

$$D_\lambda R_{\nu\sigma} = \partial_\lambda R_{\nu\sigma} - \Gamma^\omega_{\lambda\sigma} R_{\nu\omega} - \Gamma^\omega_{\lambda\nu} R_{\omega\sigma}. \tag{4.71}$$

Note that it is not implied that $R_{\nu\sigma}$ is symmetric under $\nu \leftrightarrow \nu$, so that the ordering of the indices in $R_{\nu\sigma}$ does matter. This is the covariant derivative of a tensor of rank $\binom{0}{2}$. Its components transform as those of a tensor of rank $\begin{smallmatrix}0\\3\end{smallmatrix}$.

4.9.4 Higher Ranks

You might be able to see a pattern emerging here. Every lower index that appears gets a Christoffel symbol term with a minus sign in its covariant derivative, and every upper index symbol gets a Christoffel symbol term with a plus sign. So, here is a summary of the covariant derivatives for tensors up to and including rank two (either upper, lower, or mixed):

$$D_\alpha A^\mu = \partial_\alpha A^\mu + \Gamma^\mu_{\sigma\alpha} A^\sigma$$
$$D_\alpha A_\mu = \partial_\alpha A_\mu - \Gamma^\sigma_{\alpha\mu} A_\sigma$$
$$D_\alpha B_{\beta\gamma} = \partial_\alpha B_{\beta\gamma} - \Gamma^\omega_{\alpha\gamma} B_{\beta\omega} - \Gamma^\omega_{\alpha\beta} B_{\omega\gamma}$$
$$D_\alpha B_\beta{}^\gamma = \partial_\alpha B_\beta{}^\gamma - \Gamma^\omega_{\alpha\beta} B_\omega{}^\gamma + \Gamma^\gamma_{\alpha\omega} B_\beta{}^\omega$$
$$D_\alpha B^{\beta\gamma} = \partial_\alpha B^{\beta\gamma} + \Gamma^\beta_{\alpha\omega} B^{\omega\gamma} + \Gamma^\gamma_{\alpha\omega} B^{\beta\omega}. \tag{4.72}$$

All these covariant derivatives are the components of tensors, and their ranks can be deduced by counting the upstairs and downstairs indices in their components. We will see in the next section that the Christoffel symbols are symmetric with respect to interchange of their lower two indices, so there is no need to take care over exact position of indices in Christoffel symbols.

The general rule is that every upstairs index in the tensor component gets a term added to the covariant derivative where that upstairs index is replaced by a summed index, repeated in one of the two lower indices of the Christoffel symbol, where the other lower index of the Christoffel symbol is the index of the derivative, and the upstairs index on the component being differentiated moves to the upstairs Christoffel symbol index. Similarly, every downstairs index in the tensor component gets a term subtracted from the covariant derivative where that downstairs index is replaced by a summed index, repeated in the upstairs index of the Christoffel symbol, and the two downstairs indices of the Christoffel symbol are the index of the derivative and the index replaced by the sum. Where the tensor component has multiple indices, there is one of these Christoffel symbol terms for every index, so that the expressions for Covariant derivatives of the components get more intricate as the rank of the tensor increases.

The order of the indices in the tensors of rank 2 or higher is important, which is why I took care to insert enough space in particular in the tensor $B_\beta{}^\gamma$.

4.10 The Metric Coefficients

It turns out that the coefficients of the metric are a rather special case of a tensor of rank $\binom{0}{2}$, so let us discuss them in detail. Start by writing down the metric coefficients in flat space-time and taking a partial derivative of these coefficients, $\partial_\lambda \eta_{\mu\nu}$. Since the elements $\eta_{\mu\nu}$ are all constants, it follows that

$$\partial_\lambda \eta_{\mu\nu} = 0. \tag{4.73}$$

Rewrite this equation in terms of tensors by replacing $\eta_{\mu\nu}$ with $g_{\mu\nu}$ and the ordinary derivative with a covariant derivative:

$$D_\lambda g_{\mu\nu} = 0. \tag{4.74}$$

Written entirely in terms of tensors, this equation is true in any coordinate system. So, for the particular case of the coefficients of the metric, the covariant derivative is zero. Then substituting this into Equation (4.71), we obtain

$$D_\lambda g_{v\sigma} = 0 = \partial_\lambda g_{v\sigma} - \Gamma^\omega_{\lambda\sigma} g_{v\omega} - \Gamma^\omega_{\lambda v} g_{\omega\sigma}$$
$$\partial_\lambda g_{v\sigma} = \Gamma^\omega_{\lambda\sigma} g_{v\omega} + \Gamma^\omega_{\lambda v} g_{\omega\sigma}. \tag{4.75}$$

Now let us write this equation along with its two equivalent cousins, obtained in the first case by starting with the last line of Equation (4.75) and interchanging the indices $\sigma \leftrightarrow \lambda$ and in the second case by starting again with Equation (4.75) and interchanging the indices $v \leftrightarrow \lambda$. The three equations we obtain are

$$\partial_\lambda g_{v\sigma} = \Gamma^\omega_{\lambda\sigma} g_{v\omega} + \Gamma^\omega_{\lambda v} g_{\omega\sigma};$$
$$\partial_\sigma g_{v\lambda} = \Gamma^\omega_{\sigma\lambda} g_{v\omega} + \Gamma^\omega_{\sigma v} g_{\omega\lambda};$$
$$\partial_v g_{\lambda\sigma} = \Gamma^\omega_{v\sigma} g_{\lambda\omega} + \Gamma^\omega_{v\lambda} g_{\omega\sigma}. \tag{4.76}$$

Recalling from Section 4.8.1 that the Christoffel symbols are symmetric with respect to the interchange of their two lower indices and that the metric coefficients are also symmetric, we see that the first term on the right-hand side of the third equation is equal to the second term on the right-hand side of the second equation, and the second term on the right-hand side of the third equation is equal to the second term on the right-hand side of the first equation. Notice also that the first terms on the right of the first two equations are the same as each other. Therefore if we add the first and second equations (left- and right-hand sides) and subtract the third equation, we are left with

$$\partial_\lambda g_{\nu\sigma} + \partial_\sigma g_{\nu\lambda} - \partial_\nu g_{\lambda\sigma} = 2\Gamma^\omega_{\lambda\sigma} g_{\nu\omega}. \tag{4.77}$$

We now manipulate Equation (4.77) further. Exchange the left- and right-hand sides of this equation and divide by two:

$$\Gamma^\omega_{\lambda\sigma} g_{\nu\omega} = \frac{1}{2}(\partial_\lambda g_{\nu\sigma} + \partial_\sigma g_{\nu\lambda} - \partial_\nu g_{\lambda\sigma}). \tag{4.78}$$

This looks like an equation for the Christoffel symbols in terms of ordinary derivatives of the metric components. In fact, all we have to do is to multiply by $g^{\phi\nu}$ on both sides, and this exactly becomes

$$g^{\phi\nu}\Gamma^\omega_{\lambda\sigma} g_{\nu\omega} = \frac{g^{\phi\nu}}{2}(\partial_\lambda g_{\nu\sigma} + \partial_\sigma g_{\nu\lambda} - \partial_\nu g_{\lambda\sigma})$$

$$\delta^\phi_\omega \Gamma^\omega_{\lambda\sigma} = \frac{g^{\phi\nu}}{2}(\partial_\lambda g_{\nu\sigma} + \partial_\sigma g_{\nu\lambda} - \partial_\nu g_{\lambda\sigma})$$

$$\Gamma^\phi_{\lambda\sigma} = \frac{g^{\phi\nu}}{2}(\partial_\lambda g_{\nu\sigma} + \partial_\sigma g_{\nu\lambda} - \partial_\nu g_{\lambda\sigma}). \tag{4.79}$$

This is the first expression allowing us to compute the Christoffel symbols knowing the metric coefficients. They can be found by taking ordinary derivatives of metric coefficients with respect to the coordinates of displacement. It is not a tensor expression, but that is expected because neither are the Christoffel symbols.

The Christoffel symbols appear any time you are trying to connect neighbouring points in a space or space-time. Therefore a natural place you would expect them to come up is when you work out the world line of a particle, or the geodesic in a space. After all, these lines stretch long distances across the space or space-time, but any two close-by points on the trajectory are connected to each other by a small displacement. Indeed, you can write the equation for a geodesic in terms of Christoffel symbols, and we will derive this expression in the next section.

4.11 The Geodesic Equations

Using the rules for differentiating tensors, we can work out the equation of motion for the components of the position of a particle falling freely in an arbitrary gravitational field. These so-called geodesic equations are useful, particularly for comparing the predictions of general relativity with those of Newtonian gravity. We will need to do this in Chapters 6 and 7.

The most compact way of deriving the geodesic equations is to start in an inertial frame of reference, the rest frame of an observer in free fall watching the body in question pass by. Because in this frame, there is no local evidence of the gravitational field, the velocity vector of the particle is constant – it moves in a straight line. Recall the argument in Chapter 1 about two observers playing catch in a freely falling lift, where we argued that the ball moves in a straight line at constant velocity. If $\vec{u}$ is the velocity vector of the particle, and τ is the proper time, then the velocity vector of the particle being a constant means that

$$\frac{d\vec{u}}{d\tau} = 0. \tag{4.80}$$

This is a tensor statement because both $\vec{u}$ and τ are Lorentz scalars. Therefore it is true in all coordinate systems, not just in freely falling ones. We now write $\vec{u} = u^\alpha \vec{e}_\alpha$, where the coordinate system is not necessarily in free fall. By the product rule Equation (4.80) becomes

$$\frac{d}{d\tau}\left(u^\alpha \vec{e}_\alpha\right) = \vec{e}_\alpha \frac{du^\alpha}{d\tau} + u^\alpha \frac{d\vec{e}_\alpha}{d\tau} = 0. \tag{4.81}$$

The derivative of $\vec{e}_\alpha$ with respect to τ in the second term is non-zero in non-freely falling coordinates, because as the particle moves from place to place, the coordinate system, and hence the basis vectors, are changing. We therefore use the chain rule to write this derivative as follows:

$$\begin{aligned}
\frac{d\vec{e}_\alpha}{d\tau} &= \frac{\partial \vec{e}_\alpha}{\partial x^\beta} \frac{dx^\beta}{d\tau} \\
&= \Gamma^\mu_{\alpha\beta} \vec{e}_\mu u^\beta, \tag{4.82}
\end{aligned}$$

where on the last line, we have substituted the definition of the Christoffel symbol from Equation (4.56) and written the derivative $dx^\beta/d\tau$ as the component u^β of the velocity. Substituting back into Equation (4.81), we obtain

$$\vec{e}_\alpha \frac{du^\alpha}{d\tau} + u^\alpha \Gamma^\mu_{\alpha\beta} \vec{e}_\mu u^\beta = 0. \tag{4.83}$$

We swap the repeated indices α and μ in the second term:

$$\vec{e}_\alpha \frac{du^\alpha}{d\tau} + u^\mu \Gamma^\alpha_{\mu\beta} \vec{e}_\alpha u^\beta = 0. \tag{4.84}$$

We factor out the common basis vector between the two terms:

$$\vec{e}_\alpha \left(\frac{du^\alpha}{d\tau} + \Gamma^\alpha_{\mu\beta} u^\mu u^\beta\right) = 0. \tag{4.85}$$

The component quantity in brackets must therefore be zero. Finally, we write $u^\alpha = dx^\alpha/d\tau$ and obtain the geodesic equations

$$\frac{d^2 x^\alpha}{d\tau^2} + \Gamma^\alpha_{\mu\beta} \frac{dx^\mu}{d\tau} \frac{dx^\beta}{d\tau} = 0. \tag{4.86}$$

These are the equations for a particle in free fall in a gravitational field. They are called the geodesic equations, since the world lines that are their solutions are geodesics. They are second-order differential equations, and in fact they are the curved space-time generalisations of Newton's first law of motion. Consider: in flat space-time there is no curvature, so the Christoffel symbols all vanish, and you are left with the second partial derivative of position with respect to time equal to zero; in other words, there is no acceleration, and a body continues in a state of rest or continuous momentum. Or, we can start with the principle of equivalence. If you are in free fall, then space-time appears flat. Consequently, if you build a local rec-tilinear coordinate system, then the Christoffel symbols all vanish. Consequently, the geodesics of freely falling particles in that coordinate system are straight world lines through space-time. So, to you as an observer, particles behave as they would in the absence of a gravitational field. Everything is consistent.

The geodesic equations also have practical applications. If you have used the Euler–Lagrange equations to find the equations of motion for a freely falling par-ticle for a particular problem, then you can read off the Christoffel symbols for the coordinate system you are working from by comparing the coefficients of the ve-locity terms in these equations of motion with the geodesic Equations (4.86). This is often easier in practice than using Equation (4.79). See Problems 4.5 and 4.6 for some examples of this method. For some problems, you can also solve the geodesic equations, which after all are just coupled second-order differential equations, to obtain the particle trajectories directly. In practice, however, this is often difficult, and you end up using other tricks such as identifying cyclic coordinates or, in the case of massless particle trajectories, setting $ds^2 = 0$ and solving these equations, instead.

At least in principle, however, starting from the metric coefficients, you can de-duce the trajectories taken by freely falling particles. The next question will be what are the metric coefficients? To figure these out, we need to solve for them in the presence of some distribution of matter and energy. The underlying fact is that if the metric coefficients correspond to a curved space, then there must be matter and energy present. To make that connection, we need a measure of whether our space is curved in the language of tensors. This is the subject of the next chapter.

4.12 Problems

4.1   A particle confined to the x axis with constant acceleration is governed by the differential equation $d^2x/dt^2 = a$. Show that if $v = dx/dt$, then the differential equation can be written $v(dv/dx) = a$. Now integrate the dif-ferential equation with the initial condition that $v = u$ when $x = 0$ to show that when the particle has reached a position s, its velocity $v(s)$ is such that

$v(s)^2 = u^2 + 2as$. The substitution you used here can be used to perform the first integral towards the solution of the geodesic equations for space and space-time geometries corresponding to a wide variety of metrics.

4.2 In this problem, we will prove that $dL/d\tau = 0$, Equation (4.39), in a flat one-dimensional relativistic space-time, and then consider how this result generalises to other space-time coordinate systems.

(a) Write down the Lagrangian L for a free particle on a geodesic in flat one-dimensional space-time, where $(x^0, x^1) = (ct, x)$ are the lab frame coordinates, and τ is the proper time in the rest frame of the particle, which you should use as a parameter along the geodesic.

(b) Find an expression for $dL/d\tau$ in terms of L, c, and first and second derivatives of x and t with respect to τ.

(c) Show that substituting in $dt = d\tau \cosh \eta$ and $dx = c\, d\tau \sinh \eta$ is consistent with the invariance of the proper time interval $d\tau$ under Lorentz transformations.

(d) Show that $dL/d\tau = 0$.

(e) Does the result in part (d) also apply to particles in free fall in a gravitational field? Explain your answer.

(f) Does the result in part (d) also apply to particles at fixed spherical coordinates (r, θ, ϕ), where r corresponds to twice the Schwarzschild radius, relative to a non-spinning black hole at the origin? Explain your answer.

4.3 The Lagrangian for a freely propagating particle having space-time coordinates $(x^\mu) = (ct, x, y, z)$ in zero gravitational field is

$$L = \sqrt{-g_{\mu\nu}\frac{dx^\mu}{d\tau}\frac{dx^\nu}{d\tau}},$$

where $(g_{\mu\nu}) = \text{diag}(-1, 1, 1, 1)$. Identify the four conserved quantities v_α arising from this Lagrangian and the four associated quantities v^β. Using the invariant $v_\alpha v^\alpha = -c^2$, prove the energy–momentum–mass relation for the particle, $E^2 = |\vec{p}|^2 c^2 + m_0^2 c^4$, where the total particle energy is $E = \gamma m_0 c^2$, the particle momentum is $\vec{p} = \gamma m_0 \vec{v}$, m_0 is the rest mass of the particle, and $\vec{v}$ is its velocity. You may need the equation $\gamma = dt/d\tau$, where $\gamma = 1/\sqrt{1 - v^2/c^2}$.

4.4 In classical mechanics, a free particle has the total energy $E = T + V$, where T and V are the kinetic and potential energies. The usual form of the Lagrangian is $L = T - V$. Solve simultaneously to express L in terms of T and E, then write the action S as the integral over a path over a function of p, dx (an increment in x), E, and dt (a corresponding time increment). Show further that consequently the classical mechanical Lagrangian corresponds to something Lorentz invariant and that this Lorentz invariant has some relationship to something important in quantum mechanics. This is one way of understanding the quantum mechanical origins of the least action principle in classical mechanics.

4.5   Show that in N dimensions, there are $N^2(N+1)/2$ independent Christoffel symbols. Hint: recall the symmetry with respect to exchange of the two lower indices. You will need to use $\sum_{n=1}^{N} n = N(N+1)/2$. If you want to really get to the bottom of things, then you should also prove this result. It is a neat proof if you have never encountered it. Verify that your result corresponds to reality in two and three dimensions by writing down and counting all the independent symbols.

4.6   Show that the non-zero Christoffel symbols in plane polar coordinates are $\Gamma^r_{\theta\theta} = -r$ and $\Gamma^\theta_{r\theta} = \Gamma^\theta_{\theta r} = 1/r$. You can do this either by starting from the metric in plane polar coordinates from Equation (2.12) and using the formula for Christoffel symbols given in Equation (4.79), or by starting from the Lagrangian, working out the Euler–Lagrange equations, and then reading off the Christoffel symbols using Equation (4.86). I think the latter method is quicker, particularly if you want to prove that all the Christoffel symbols other than those given above are zero.

4.7   Work out the differential equations for a particle moving in a straight line in a plane, but unlike the discussion in Chapter 2, using plane polar coordinates. The Euler–Lagrange equations should lead to two coupled second-order differential equations in the parameters r and θ. These can either be worked out directly from the metric coefficients given in Equation (2.12), or if you did Problem 4.6, then you can express them directly from the geodesic equations. You should then establish a cyclic coordinate, deduce the associated conserved quantity, and use this to decouple the equations, resulting in two 'homogeneous' equations, one in θ with no r dependence and the other in r with no θ dependence.

4.8   Consider spherical polar coordinates in a three-dimensional flat space.

(a)   If you have not already done so in Problem 2.2, work out the components of the metric in spherical polar coordinates.

(b)   Work out expressions for all 18 independent Christoffel symbols in spherical polar coordinates. One approach to doing this problem is to write down the Lagrangian, derive the Euler–Lagrange equations, and by comparing with the general geodesic equations of Equation (4.86), read off both the non-zero Christoffel symbols, which will be represented in your Euler–Lagrange equations, and the non-zero Christoffel symbols, which will not be represented. Do not forget that because of the symmetry of $\Gamma^\alpha_{\beta\gamma}$ with respect to the interchange $\beta \leftrightarrow \gamma$, two such Christoffel symbols will each take half of the corresponding term that may show up in one of your equations of motion. Another approach is repeated application of Equation (4.79). In the author's opinion, the latter is more difficult.

4.9 Gauss' law for the electric field in the absence of charges is $\vec{\nabla} \cdot \vec{E} = 0$.

(a) Write Gauss' law in component notation using the operator ∂_i.

(b) Replace the non-tensor aspects of this expression with tensor equivalents that are the same as the non-tensor aspects they are replacing in Cartesian coordinates, but differ in other coordinate systems. The resulting expression is Gauss' law valid in any coordinate system. It will contain Christoffel symbols.

(c) If you have done Problem 4.8, then you can replace the Christoffel symbols with their expressions in terms of the spherical coordinates (r, θ, ϕ). Compare the resulting expression for Gauss' law with those appearing in electromagnetism textbooks, or with the results commonly quoted in formula sheets. Beware! Those results assume unit length basis vectors, so you will need to re-scale the magnitudes of components of $\vec{E}$.

4.10 Consider the gradient of a scalar field, $\vec{\nabla}V$. This is a more complex example than the $\vec{\nabla} \cdot \vec{E}$ encountered in Problem 4.9, because the quantity is a vector. In component notation, this quantity can be written $(\vec{\nabla}V)^i \vec{e}_i$.

(a) Start in Cartesian coordinates with the usual expression for $\vec{\nabla}V$,

$$\vec{\nabla}V = \frac{\partial V}{\partial x}\vec{e}_x + \frac{\partial V}{\partial y}\vec{e}_y + \frac{\partial V}{\partial z}\vec{e}_z = \frac{\partial V}{\partial x^i}\vec{e}_i.$$

This is not a tensor expression because the i indices do not balance. Remembering that we are in Cartesian coordinates, show that the gradient of V can be written

$$\vec{\nabla}V = g^{ij}\frac{\partial V}{\partial x^j}\vec{e}_i.$$

This is now a tensor expression.

(b) Work out expressions for g^{rr}, $g^{\theta\theta}$, and $g^{\phi\phi}$ in spherical polar coordinates. These follow from the metric coefficients calculated in Problem 2.2.

(c) Work out expressions for the unit vectors $\hat{e}_r$, $\hat{e}_\theta$, and $\hat{e}_\phi$ in terms of the basis vectors $\vec{e}_r$, $\vec{e}_\theta$, and $\vec{e}_\phi$.

(d) Show that the gradient of V in spherical polar coordinates can be written

$$\vec{\nabla} V = \frac{\partial V}{\partial r}\hat{e}_r + \frac{1}{r}\frac{\partial V}{\partial \theta}\hat{e}_\theta + \frac{1}{r\sin\theta}\frac{\partial V}{\partial \phi}\hat{e}_\phi.$$

This result will be familiar from formula sheets and the cheat sheets inside the covers of electromagnetism books!

4.11 Work out the Christoffel symbols on the surface of a sphere in restricted spherical polar coordinates where the origin is at the centre of the sphere and as a consequence the radius $r = R$ is a constant. I think it is less laborious to do this by working out the Euler–Lagrange equations and reading off the Christoffel symbols from the geodesic equations (4.86), though it is healthy to compute them both ways and check that they are the same.

4.12 Write down a tensor expression for $\vec{\nabla} \times \vec{E}$, where $\vec{E} = E^i \vec{e}_i$, in three dimensions. Start by writing down an expression that gives the correct answer in Cartesian coordinates, then manipulate indices in a manner that maintains the right answer, but where necessary balance indices and replace ordinary derivatives with covariant ones. You will need the Levi-Civita tensor discussed in Problems 2.8, 2.9, and 2.10.

4.13 Evaluate the radial component of $\vec{\nabla} \times \vec{E}$ in spherical polar coordinates. Note that you will need $\varepsilon^{r\theta\phi}$, which was evaluated in Problem 2.10. The answer for the radial component is given in many books. In terms of components $\vec{E} = E^r \hat{e}_r + E^\theta \hat{e}_\theta + E^\phi \hat{e}_\phi$ with unit vectors, which is the common convention, the answer is

$$(\vec{\nabla} \times \vec{E})^r = \frac{1}{r\sin\theta}\left(\frac{\partial}{\partial\theta}(\sin\theta\, E^\phi) - \frac{\partial E^\theta}{\partial\phi}\right).$$

5

Einstein's Equations

5.1 Curvature Tensors

In Chapter 4, we made the connection between the coefficients of the metric of a curved space or space-time and the geodesics there. In space, the geodesics are the paths followed by stretched rubber bands, paths of minimum length. In space-time, the geodesics are world lines of particles in free fall. In a freely falling frame, those world lines appear straight; in a non-freely falling frame, the world lines take on the curvature resulting from gravitational fields. In this chapter, we will make the connection between matter and energy in a space and the resultant gravitational field.

To do this, we will need to work out a tensor expression that tells us whether, to some observer, space-time appears flat or curved. We have seen previously that the coefficients of the metric themselves do not furnish this information, since even in flat space, curvilinear coordinate systems give rise to position-dependent and non-trivial metric coefficients. There do, however, exist tensors that tell you whether your space is flat or curved. These are the Riemann and Ricci tensors, the Ricci scalar, and finally the Einstein tensor. To find them out, we need to think about a concept called parallel transport.

5.2 Parallel Transport

If I draw an arrow on a flat sheet of paper, then there is an operation that we all know how to do, or imagine doing, called parallel transport. In parallel transport, the point at the start of the arrow is moved, or transported, to a new place. The arrow transports with the point, maintaining the same direction that it had originally, so that it remains parallel to the original arrow or vector. This is what we mean by parallel transport. You can also imagine parallel transporting a vector around a closed loop and coming back to where you started. If you do this, then the vectors at the beginning and end of the closed path should be the same. Furthermore, even if we are using curvilinear coordinates to parameterise the vector, the components of the vector may vary as we travel around the closed loop, but when we get back to the starting point, the components of the vector should be the same as they were before we started.

Next, we go to a curved space. It is not so obvious how to do parallel transport in a curved space, but we can make workable definitions of this operation where there are flat tangent spaces that we can use. As an example, consider a spherical surface. If we move around the equator, which is a great circle, then every point on that equator is tangent to a point on a cylinder wrapped around the sphere at the equator. So, if we are only interested in parallel transporting around the equator, then when we get back to where we started, our arrow will be in the same direction it was before the start of the trip. This works because we can imagine unwrapping the cylinder into a flat sheet, and the line of points that was touching the equator forms a straight line across the sheet from one side to the other. Parallel transport on the flat sheet proceeds exactly the same as already discussed for flat spaces. At the end, we pick up the sheet and wrap it back around the sphere, at which point the arrows at the beginning and end of the journey are parallel to each other, and all the origins of the arrows are in tangent planes to the equator.

A slightly more complicated case is that of a line of constant latitude on a sphere, so a circular path at constant θ, which is not 90°. There is no cylindrical shaped tangent space that touches all the points on this circle. However, if you make a circular cone of just the right opening angle, then you can drop the sphere into the cone, and it will touch it at every point on a line of constant latitude. So, the surface of the cone is tangent to the surface of the sphere on one particular line of latitude. Now, to flatten the cone onto a table, I make a cut from the point of the cone that extends outwards to the line of constant latitude and then keeps going until it reaches the end of the cone, wherever that is. When I flatten this sheet onto the table, I see that it forms the sector of a circle, with a sector of paper missing with an interior angle that depends on the latitude θ at which the cone was tangent. I can still do parallel transport of a vector along the line, but I cannot go back to where I started on the paper because of the missing sector. When the cone is bought back into contact with the sphere, the arrows on either side of the boundary are no longer parallel; there is an angle between them that is the same as the opening angle of the sector when the cone was flattened out.

A non-zero result for parallel transport of a vector around a closed loop on a surface is an indicator that the space was curved for at least some of the pathway along which parallel transport was attempted.

We will shortly discover a tensor that tells you the result for parallel transport of a vector around a small closed loop in a space. In fact, it tells you what will happen when you do parallel transport of any vector around closed loops that are aligned with respect to any pair of coordinates in the space. Our small closed loop will consist first of a small displacement δx^α in the direction of basis vector $\vec{e}_\alpha$, then from the end point of that displacement a further small displacement δx^β in the direction of basis vector $\vec{e}_\beta$. We could instead have carried out the small displacements in reverse order, in which case we will end up at the same final point, but we will have got there by taking a different pathway. In general, the directions of the vectors from parallel transport by the two paths will be different. Differences between the vectors after parallel transport by these two routes indicate that the space is curved in the vicinity of the small loop.

To do parallel transport, we need tangent planes to the points between which we are moving. Figure 5.1 shows how this is possible. There are four points you are moving between when parallel transporting between points P and Q by the two different routes. Each point has a different tangent plane, but as before, these tangent planes intersect at straight lines, so we can define parallel transport between the four points as the operation that gives the same outcome as parallel transport between the points in the tangent planes. Start from point P, which has position $\vec{r}$. At this point, you have a vector $\vec{A}$ with components A^μ. Now move to a neighbouring point $\vec{r} + \delta x^\alpha \vec{e}_\alpha$. If we were in a flat space, then the components of $\vec{A}$ would undergo a modification

$$A^\mu(\vec{r} + \delta x^\alpha \vec{e}_\alpha) = A^\mu(\vec{r}) + \frac{\partial A^\mu}{\partial x^\alpha} \delta x^\alpha. \tag{5.1}$$

Next, we move in the direction of $\vec{e}_\beta$. Were we in a flat space, the vector components after the two shifts, $A^\mu(\vec{r} + \delta x^\alpha \vec{e}_\alpha + \delta x^\beta \vec{e}_\beta)$, would be

$$A^\mu(\vec{r} + \delta x^\alpha \vec{e}_\alpha + \delta x^\beta \vec{e}_\beta) = A^\mu(\vec{r}) + \frac{\partial A^\mu}{\partial x^\alpha} \delta x^\alpha + \frac{\partial}{\partial x^\beta}\left(A^\mu(\vec{r}) + \frac{\partial A^\mu}{\partial x^\alpha} \delta x^\alpha\right)\delta x^\beta$$

$$= A^\mu(\vec{r}) + \frac{\partial A^\mu}{\partial x^\alpha} \delta x^\alpha + \frac{\partial A^\mu}{\partial x^\beta} \delta x^\beta$$

$$+ \frac{\partial}{\partial x^\beta} \frac{\partial}{\partial x^\alpha} A^\mu(\vec{r}) \delta x^\alpha \delta x^\beta. \tag{5.2}$$

Now we work out the correct expressions in a curved space by replacing the non-tensor objects in Equation (5.2), which are the partial derivatives, with covariant derivatives:

$$A^\mu(\vec{r} + \delta x^\alpha \vec{e}_\alpha + \delta x^\beta \vec{e}_\beta)$$
$$= A^\mu(\vec{r}) + D_\alpha A^\mu \delta x^\alpha + D_\beta A^\mu \delta x^\beta + D_\beta D_\alpha A^\mu \delta x^\alpha \delta x^\beta. \tag{5.3}$$

This is the correct expression for the evolution of the components of $\vec{A}$ under parallel transport from P to Q by the first route along $\vec{e}_\alpha$ followed by $\vec{e}_\beta$. We now figure out the outcome of parallel transport to the same point Q but this time first

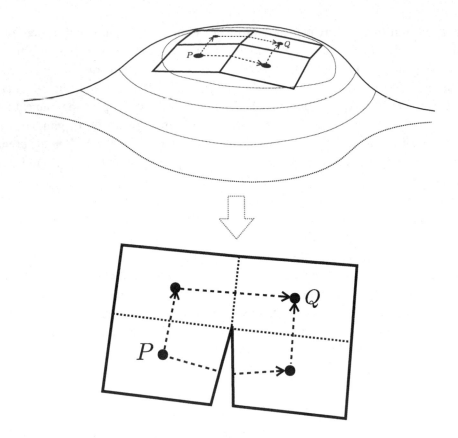

Figure 5.1 Parallel transport between points P and Q in a curved surface by two different routes, showing the tangent planes at the end points of the different portions of the motion and how these tangent planes intersect at straight lines. As with the cone on a sphere, there is a missing sector when the assembly of tangent planes is laid out flat. The opposite edges of the sector join together when the tangent plane assembly is replaced on the curved surface.

travelling in the direction $\vec{e}_\beta$ and subsequently travelling in the direction $\vec{e}_\alpha$:

$$A^\mu_{\text{ALT}}\left(\vec{r} + \delta x^\alpha \vec{e}_\alpha + \delta x^\beta \vec{e}_\beta\right)$$
$$= A^\mu(\vec{r}) + D_\alpha A^\mu \delta x^\alpha + D_\beta A^\mu \delta x^\beta + D_\alpha D_\beta A^\mu \delta x^\alpha \delta x^\beta. \qquad (5.4)$$

The difference between the components A^μ and A^μ_{ALT} is given by the only terms in Equations (5.3) and (5.4) that do not commute, and that is the covariant derivatives:

$$A^\mu - A^\mu_{\text{ALT}} = \left(D_\beta D_\alpha A^\mu - D_\alpha D_\beta A^\mu\right)\delta x^\alpha \delta x^\beta. \qquad (5.5)$$

We follow what we did before and express the difference between the vector components before and after the parallel transport by these two routes as linear superpositions of the components at the starting point P:

$$(D_\beta D_\alpha - D_\alpha D_\beta)A^\mu = R^\mu{}_{\kappa\beta\alpha}A^\kappa. \tag{5.6}$$

If we were in a flat space, then we could transform to rectilinear coordinates, and the covariant derivatives are then just equivalent to partial derivatives, which then commute, so the left-hand side is zero, and consequently we would conclude that $R^\mu{}_{\kappa\beta\alpha} = 0$. Because all the other quantities in Equation (5.6) are tensors, $R^\mu{}_{\kappa\beta\alpha}$ is a tensor too, of rank $\frac{1}{3}$. It is the Riemann curvature tensor.

5.3 Curvature from Christoffel Symbols

We can express the Riemann curvature tensor in terms of Christoffel symbols and their spatial derivatives. Consider first $D_\beta D_\alpha A^\mu = D_\beta T_\alpha{}^\mu$. This is the covariant derivative of a tensor of rank $\frac{1}{1}$:

$$\begin{aligned}
D_\beta T_\alpha{}^\mu &= \partial_\beta T_\alpha^\mu + \Gamma^\mu_{\beta\kappa} T_\alpha^\kappa - \Gamma^\lambda_{\beta\alpha} T_\lambda^\mu \\
&= \partial_\beta D_\alpha A^\mu + \Gamma^\mu_{\beta\kappa} D_\alpha A^\kappa - \Gamma^\lambda_{\beta\alpha} D_\lambda A^\mu \\
&= \partial_\beta \partial_\alpha A^\mu + \partial_\beta \Gamma^\mu_{\alpha\lambda} A^\lambda + \Gamma^\mu_{\beta\kappa} \partial_\alpha A^\kappa + \Gamma^\mu_{\beta\kappa} \Gamma^\kappa_{\alpha\lambda} A^\lambda \\
&\quad - \Gamma^\lambda_{\beta\alpha} \partial_\lambda A^\mu - \Gamma^\lambda_{\beta\alpha} \Gamma^\mu_{\lambda\kappa} A^\kappa.
\end{aligned} \tag{5.7}$$

The second term $D_\alpha T_\beta{}^\mu$ is the same thing with α and β interchanged, so we just swap these symbols everywhere compared to the last line in the previous equation:

$$\begin{aligned}
D_\alpha T_\beta{}^\mu &= \partial_\alpha \partial_\beta A^\mu + \partial_\alpha \Gamma^\mu_{\beta\lambda} A^\lambda + \Gamma^\mu_{\alpha\kappa} \partial_\beta A^\kappa + \Gamma^\mu_{\alpha\kappa} \Gamma^\kappa_{\beta\lambda} A^\lambda \\
&\quad - \Gamma^\lambda_{\alpha\beta} \partial_\lambda A^\mu - \Gamma^\lambda_{\alpha\beta} \Gamma^\mu_{\lambda\kappa} A^\kappa.
\end{aligned} \tag{5.8}$$

The first terms and the last two terms in each expression cancel when you take the difference,

$$\begin{aligned}
(D_\beta D_\alpha - D_\alpha D_\beta)A^\mu &= \partial_\beta \Gamma^\mu_{\alpha\lambda} A^\lambda + \Gamma^\mu_{\beta\kappa} \partial_\alpha A^\kappa + \Gamma^\mu_{\beta\kappa} \Gamma^\kappa_{\alpha\lambda} A^\lambda \\
&\quad - \partial_\alpha \Gamma^\mu_{\beta\lambda} A^\lambda - \Gamma^\mu_{\alpha\kappa} \partial_\beta A^\kappa - \Gamma^\mu_{\alpha\kappa} \Gamma^\kappa_{\beta\lambda} A^\lambda.
\end{aligned} \tag{5.9}$$

The first term is expanded by the product rule to $A^\lambda \partial_\beta \Gamma^\mu_{\alpha\lambda} + \Gamma^\mu_{\alpha\lambda} \partial_\beta A^\lambda$, but the second term in this expansion cancels with the fifth term. Similarly, the fourth term is $-A^\lambda \partial_\alpha \Gamma^\mu_{\beta\lambda} - \Gamma^\mu_{\beta\lambda} \partial_\alpha A^\lambda$, but the second term in this expansion cancels with the second term. So, we are left with

$$(D_\beta D_\alpha - D_\alpha D_\beta)A^\mu = A^\lambda \partial_\beta \Gamma^\mu_{\alpha\lambda} + \Gamma^\mu_{\beta\kappa} \Gamma^\kappa_{\alpha\lambda} A^\lambda - A^\lambda \partial_\alpha \Gamma^\mu_{\beta\lambda} - \Gamma^\mu_{\alpha\kappa} \Gamma^\kappa_{\beta\lambda} A^\lambda. \tag{5.10}$$

Factoring out the A^λ, which appears in all the terms, we arrive at

$$\begin{aligned}
(D_\beta D_\alpha - D_\alpha D_\beta)A^\mu &= \left(\partial_\beta \Gamma^\mu_{\alpha\lambda} + \Gamma^\mu_{\beta\kappa} \Gamma^\kappa_{\alpha\lambda} - \partial_\alpha \Gamma^\mu_{\beta\lambda} - \Gamma^\mu_{\alpha\kappa} \Gamma^\kappa_{\beta\lambda}\right) A^\lambda \\
&= R^\mu{}_{\lambda\beta\alpha} A^\lambda,
\end{aligned} \tag{5.11}$$

where

$$R^{\mu}{}_{\lambda\beta\alpha} = \partial_{\beta}\Gamma^{\mu}_{\alpha\lambda} - \partial_{\alpha}\Gamma^{\mu}_{\beta\lambda} + \Gamma^{\mu}_{\beta\kappa}\Gamma^{\kappa}_{\alpha\lambda} - \Gamma^{\mu}_{\alpha\kappa}\Gamma^{\kappa}_{\beta\lambda}. \tag{5.12}$$

This expression allows the components of the Riemann curvature to be calculated from the Christoffel symbols. In turn, the Christoffel symbols are calculated from the components of the first derivatives of the metric and the components of the metric inverse.

In our ordinary four-dimensional space-time, there are $4^4 = 256$ Riemann tensor coefficients. Given some metric, evaluation of such a huge number of coefficients using Equation (5.12) is a formidable task. Fortunately, it turns out that many of the coefficients can be shown to be zero, and of the remaining coefficients, many are related to each other by symmetries. In the next section, we work out the symmetries of the Riemann tensor, and in the section after that we prove, quite generally, that only twenty of these Riemann coefficients need to be calculated to deduce all the others. This is an enormous simplification, which will allow us to study in some detail three problems of great interest in the final three chapters of this book. Though the mathematics is again lengthy and at times tedious, the author encourages the interested student to take courage and go through as much of it as possible. Of course, it is also possible to skip these sections rather superficially and use the results, and this is also a pathway to progress, albeit a less satisfying one. Besides, the algebra will do you good and toughen you up!

5.4 Symmetries of the Riemann Tensor

It will turn out that there are five symmetries relating different components of the Riemann curvature tensor. The first of these follows from Equation (5.12) by inspection. If we perform the interchange $\alpha \rightarrow \beta$, then the first and second terms on the right, and the third and fourth terms on the right, both interchange, leading to an expression with an overall minus sign compared to before the interchange. Therefore

$$R^{\mu}{}_{\lambda\alpha\beta} = -R^{\mu}{}_{\lambda\beta\alpha}. \tag{5.13}$$

The remaining symmetries are obscure in Equation (5.12). However, we can write the Riemann curvature in a form where several of the other symmetries can be seen.

This form is valid in the coordinates of the locally coincident, or 'pigeon' observers discussed in Section 1.6. Recall that these observers correspond, in space-time, to the frames of reference of observers comoving with, and sharing a common origin of coordinates with, a freely falling frame of reference. For such observers, the first derivatives of the metric coefficients with respect to displacements are zero, as shown first in Equation (1.13) in two spatial dimensions.

However, the Christoffel symbols themselves are proportional to first derivatives of $g_{\mu\nu}$, and therefore the last two terms in Equation (5.12) can be neglected in these coordinate systems. This leaves a reduced expression for the Riemann curvature tensor,

$$R^{\mu}{}_{\lambda\beta\alpha} = \partial_\beta \Gamma^{\mu}_{\alpha\lambda} - \partial_\alpha \Gamma^{\mu}_{\beta\lambda}. \tag{5.14}$$

We next lower the upstairs index:

$$\begin{aligned}
R_{\phi\lambda\beta\alpha} &= g_{\phi\mu} R^{\mu}{}_{\lambda\beta\alpha} \\
&= g_{\phi\mu} \partial_\beta \Gamma^{\mu}_{\alpha\lambda} - g_{\phi\mu} \partial_\alpha \Gamma^{\mu}_{\beta\lambda}. \tag{5.15}
\end{aligned}$$

We now substitute in the definition of the Christoffel symbols in terms of first derivatives of the metric coefficients from Equation (4.79):

$$\begin{aligned}
R_{\phi\lambda\beta\alpha} &= g_{\phi\mu} \partial_\beta \frac{g^{\mu\omega}}{2} (\partial_\alpha g_{\lambda\omega} + \partial_\lambda g_{\alpha\omega} - \partial_\omega g_{\alpha\lambda}) \\
&\quad - g_{\phi\mu} \partial_\alpha \frac{g^{\mu\omega}}{2} (\partial_\beta g_{\lambda\omega} + \partial_\lambda g_{\beta\omega} - \partial_\omega g_{\beta\lambda}). \tag{5.16}
\end{aligned}$$

The only non-zero terms are going to be those that contain second derivatives of metric coefficients, so we can neglect the terms where the second partial derivative operates on $g^{\mu\phi}$. We are therefore left with

$$\begin{aligned}
R_{\phi\lambda\beta\alpha} &= g_{\phi\mu} \frac{g^{\mu\omega}}{2} \partial_\beta (\partial_\alpha g_{\lambda\omega} + \partial_\lambda g_{\alpha\omega} - \partial_\omega g_{\alpha\lambda}) \\
&\quad - g_{\phi\mu} \frac{g^{\mu\omega}}{2} \partial_\alpha (\partial_\beta g_{\lambda\omega} + \partial_\lambda g_{\beta\omega} - \partial_\omega g_{\beta\lambda}) \\
&= \frac{\delta^{\omega}_{\phi}}{2} \partial_\beta (\partial_\alpha g_{\lambda\omega} + \partial_\lambda g_{\alpha\omega} - \partial_\omega g_{\alpha\lambda}) \\
&\quad - \frac{\delta^{\omega}_{\phi}}{2} \partial_\alpha (\partial_\beta g_{\lambda\omega} + \partial_\lambda g_{\beta\omega} - \partial_\omega g_{\beta\lambda}) \\
&= \frac{1}{2} (\partial_\beta \partial_\alpha g_{\lambda\phi} + \partial_\beta \partial_\lambda g_{\alpha\phi} - \partial_\beta \partial_\phi g_{\alpha\lambda} \tag{5.17} \\
&\quad - \partial_\alpha \partial_\beta g_{\lambda\phi} - \partial_\alpha \partial_\lambda g_{\beta\phi} + \partial_\alpha \partial_\phi g_{\beta\lambda}). \tag{5.18}
\end{aligned}$$

The first and fourth terms cancel, so we are led to

$$R_{\phi\lambda\beta\alpha} = \frac{1}{2} (\partial_\beta \partial_\lambda g_{\alpha\phi} - \partial_\beta \partial_\phi g_{\alpha\lambda} - \partial_\alpha \partial_\lambda g_{\beta\phi} + \partial_\alpha \partial_\phi g_{\beta\lambda}). \tag{5.19}$$

In this form, you can see some more of the symmetries of the Riemann tensor. From before we already have

$$R_{\phi\lambda\alpha\beta} = -R_{\phi\lambda\beta\alpha}. \tag{5.20}$$

Starting with Equation (5.19), if we make the exchange $\phi \leftrightarrow \lambda$, this swaps the first and second terms and also the third and fourth terms, introducing an overall minus sign. Therefore

$$R_{\lambda\phi\beta\alpha} = -R_{\phi\lambda\beta\alpha}. \tag{5.21}$$

If, on the other hand, we do $\alpha \leftrightarrow \phi$ and $\beta \leftrightarrow \lambda$, then this exchanges the middle two terms and leaves the end two terms the same, so that

$$R_{\alpha\beta\lambda\phi} = +R_{\phi\lambda\beta\alpha}. \tag{5.22}$$

If we make the exchanges $\beta \leftrightarrow \phi$ and $\alpha \leftrightarrow \lambda$, then we exchange the first and last terms and leave the middle two the same, so

$$R_{\beta\alpha\phi\lambda} = +R_{\phi\lambda\beta\alpha}. \tag{5.23}$$

The final symmetry comes from summing together the latter three indices in $R_{\phi\lambda\alpha\beta}$ and adding together the three copies:

$$
\begin{aligned}
R_{\phi\lambda\beta\alpha} + R_{\phi\beta\alpha\lambda} + R_{\phi\alpha\lambda\beta} = \frac{1}{2}(&\partial_\beta\partial_\lambda g_{\alpha\phi} - \partial_\beta\partial_\phi g_{\alpha\lambda} - \partial_\alpha\partial_\lambda g_{\beta\phi} + \partial_\alpha\partial_\phi g_{\beta\lambda} \\
&+ \partial_\alpha\partial_\beta g_{\lambda\phi} - \partial_\alpha\partial_\phi g_{\lambda\beta} - \partial_\lambda\partial_\beta g_{\alpha\phi} + \partial_\lambda\partial_\phi g_{\alpha\beta} \\
&+ \partial_\lambda\partial_\alpha g_{\beta\phi} - \partial_\lambda\partial_\phi g_{\beta\alpha} - \partial_\beta\partial_\alpha g_{\lambda\phi} + \partial_\beta\partial_\phi g_{\lambda\alpha}) \\
= 0.&
\end{aligned} \tag{5.24}
$$

Equations (5.13), (5.21), (5.22), (5.23), and (5.24) are the complete set of symmetries of the Riemann curvature. You may object that we have found symmetries only of a related object $R_{\phi\lambda\beta\alpha}$, with all its indices lowered, and that we have in any case only used an expression for this object for a restricted set of 'pigeon' observers. However, for the problems we will consider in the remainder of this book, the metric coefficients are diagonal, so that raising and lowering indices does not result in additional terms. Even in more complicated geometries, such as the Kerr geometry, where there are off-diagonal metric terms, the raising and lowering of indices still allows us to start with $R^\mu{}_{\lambda\beta\alpha}$, lower the first index, then exploit the symmetry in the resulting terms, and finally raise the indices to return to the mixed tensor components. It is just that you end up with more terms as the lowering or raising with the metric results in a superposition. It is more complicated but conceptually the same.

As for the pigeon frames, we can invoke the principle of equivalence to generalise any symmetry of the Riemann tensor coefficients established for pigeon observers, since any of the equations above are tensor equations, so that you can reach the coordinate system of any other non-pigeon observer by applying a coordinate transformation, obtaining the same equation in the new primed coordinate system. As usual, these transformations can either be Lorentz boosts, rotations, or changes of coordinates to some convenient curvilinear coordinate basis; all these transformations leave the symmetries looking exactly the same to any non-pigeon observer.

5.4.1 Zero Riemann Components

We first identify the coefficients of the Riemann curvature that are zero for all space-time geometries. We work with $R_{\alpha\beta\gamma\delta}$, the all-index-lowered version of the Riemann curvature for which we have derived the symmetries.

We divide the Riemann curvature coefficients into 10 classes, as shown in Table 5.1. Classes A, B, and C all have either the first two indices equal or the last two indices equal. However, according to Section 5.4, the Riemann tensor is anti-symmetric with respect to interchange of either of these pairs of indices, implying that each of these components is equal to minus itself and therefore can only be zero. This means that a total of 112 of the 256 components of the Riemann tensor are zero, leaving the remaining 144 as potentially non-zero.

Table 5.1 *The ten classes of Riemann coefficients. The first column is a letter name for the class. The second column is the component, showing which indices are constrained to be equal, constrained to be different, and unconstrained. The third column is the number of possible values of each index. The fourth column indicates whether the components in this class have to be zero because of antisymmetries. The fifth column is the number of tensor components in the class, and the sixth column is the cumulative sum of the numbers in the fifth column, adding up to $2^4 = 256$, which shows that the classes cover all the components. Classes A, B, and C are always zero because of the two antisymmetries of the tensor. This is discussed in Section 5.4.1. Classes D through J are in general non-zero, but many of them are related to each other by the three symmetries of the tensor. This is discussed in Section 5.4.2.*

A	$R_{\alpha\alpha\beta\{\gamma\neq\beta\}}$	$\alpha:4,\ \beta:4,\ \gamma:3$	$=0$	48	48
B	$R_{\alpha\{\beta\neq\alpha\}\gamma\gamma}$	$\alpha:4,\ \beta:3,\ \gamma:4$	$=0$	48	96
C	$R_{\alpha\alpha\beta\beta}$	$\alpha:4,\ \beta:4$	$=0$	16	112
D	$R_{\alpha\{\beta\neq\alpha\}\alpha\{\beta\neq\alpha\}}$	$\alpha:4,\ \beta:3$	$\neq0$	12	124
E	$R_{\alpha\{\beta\neq\alpha\}\{\beta\neq\alpha\}\alpha}$	$\alpha:4,\ \beta:3$	$\neq0$	12	136
F	$R_{\alpha\{\beta\neq\alpha\}\{\gamma\neq\beta\neq\alpha\}\{\delta\neq\gamma\neq\beta\neq\alpha\}}$	$\alpha:4,\ \beta:3,\ \gamma:2,\ \delta:1$	$\neq0$	24	160
G	$R_{\alpha\{\beta\neq\alpha\}\alpha\{\gamma\neq\beta\neq\alpha\}}$	$\alpha:4,\ \beta:3,\ \gamma:2$	$\neq0$	24	184
H	$R_{\{\beta\neq\alpha\}\alpha\{\gamma\neq\beta\neq\alpha\}\alpha}$	$\alpha:4,\ \beta:3,\ \gamma:2$	$\neq0$	24	208
I	$R_{\alpha\{\beta\neq\alpha\}\{\gamma\neq\beta\neq\alpha\}\alpha}$	$\alpha:4,\ \beta:3,\ \gamma:2$	$\neq0$	24	232
J	$R_{\{\beta\neq\alpha\}\alpha\alpha\{\gamma\neq\beta\neq\alpha\}}$	$\alpha:4,\ \beta:3,\ \gamma:2$	$\neq0$	24	256

5.4.2 Independent Riemann Components

Class D

These components are of the form $R_{\alpha\{\beta\neq\alpha\}\alpha\{\beta\neq\alpha\}}$, so we write out all 12 possibilities: R_{0101}, R_{0202}, R_{0303}, R_{1010}, R_{1212}, R_{1313}, R_{2020}, R_{2121}, R_{2323}, R_{3030}, R_{3131}, R_{3232}. Exactly half of these have $\alpha < \beta$, and the other half have $\alpha > \beta$. But a symmetry of the Riemann tensors discussed in Section 5.4 is that they are symmetric under simultaneous interchange of the first two indices and the last two indices. Therefore the independent components in class D can be written $R_{\alpha\{\beta>\alpha\}\alpha\{\beta>\alpha\}}$. The other six are not zero, but they are each equal to one of the following six: $\mathbf{R_{0101}}$, $\mathbf{R_{0202}}$, $\mathbf{R_{0303}}$, $\mathbf{R_{1212}}$, $\mathbf{R_{1313}}$, and $\mathbf{R_{2323}}$.

Next, we work out the consequences of the two remaining symmetries from Section 5.4. The first involves interchange of the first and third indices simultaneous with interchange of the second and fourth indices, but for class D, these pairs of indices are equal anyway. The final symmetry $R_{\alpha[\beta\gamma\delta]} = 0$, meaning that the sum of the three symbols with the latter three indices cyclically commuted is zero, leads to expressions like $R_{0101} + R_{0110} + R_{0011} = 0$. The third term is zero because it is in class C, so we conclude that $R_{0101} = -R_{0110}$. We knew this anyway because of the antisymmetry of all the symbols with respect to the interchange of the latter two indices. This does not have any consequences for class D because R_{0110} is in class E, but it will have consequences for class E as we will see below.

This exhausts all symmetries that might further constrain the coefficients in class D, so that the six in boldface will have to be evaluated separately.

Class E

As for class D, we consider subclasses $R_{\alpha\{\beta>\alpha\}\{\beta>\alpha\}\alpha}$ and $R_{\alpha\{\beta<\alpha\}\{\beta<\alpha\}\alpha}$ separately. Each coefficient of the former subclass is equal to a coefficient of the latter subclass by the symmetry under the simultaneous interchange of the first and last pairs of indices. This leaves six independent coefficients in class E, but these are each equal to minus a component from class D by the symmetry $R_{\alpha[\beta\gamma\delta]} = 0$ discussed in Section 5.4.2. Consequently, there are no independent Riemann coefficients in class E.

Class F

These are all the coefficients where each of the indices takes a different value. We go through the possibilities to discover all the coefficients. If the first index is 0, then the second one can be either 1, 2, or 3. If the second index is 1, then the third and fourth indices can either be 23 or 32. So overall, we have 0123 or 0132 for the indices. Notice, however, that these are equal to minus each other because of antisymmetry with respect to interchange of the last two components. Furthermore, we also have indices 0123 equal to 1023 and 0132 equal to 1032 by antisymmetry with respect to interchange of the first two components. So R_{0123}, R_{0132}, R_{1023}, and R_{1032} are all related by symmetries and antisymmetries. Now we go to the first two indices being 02. By the same argument we have $R_{0213} = -R_{0231} = -R_{2013} = R_{2031}$ and similarly for the first two indices being 03. Therefore, for each combination of the first two indices, we get a set of four interrelated coefficients. There are 24 coefficients in class F altogether, so they group into six groups of four. For the six groups, the first two indices are, in the six cases: 01, 02, 03, 12, 13, 23. So, there are at most six independent coefficients in class F.

However, we still have to utilise the other two symmetries, the first where you interchange the first and third indices simultaneously with the second and fourth indices. For example, $R_{0123} = R_{2301}$. This means that the class with the first two indices 01 are not independent of the class with the first two indices 23. Similarly, starting with R_{0213} and R_{0312}, we can show that the classes with 02 and 03 as their first indices are connected to the classes with 13 and 12 as their first indices, respectively. This means, then, that at most three coefficients are independent – for example, R_{0123}, R_{0213}, and R_{0312}. Now there is one more symmetry to exploit, that where $R_{\alpha[\beta\gamma\delta]} = 0$. This means, for example, that $R_{0123} + R_{0312} + R_{0231} = 0$. This implies that if we know a member of the class whose first two indices are 01, and a member of the class whose first two indices are 02, then we can deduce the value of a member whose first two indices are 03. Thus there are in fact only two independent coefficients in class F. We will take them as $\mathbf{R_{0123}}$ and $\mathbf{R_{0213}}$.

Class G

Class G has 24 members of the form $R_{\alpha\beta\alpha\gamma}$. Because of the symmetry under simultaneous interchange of the first and third indices and the second and fourth indices, the 12 coefficients where $\beta > \gamma$ are equal to the 12 coefficients where $\beta < \gamma$, so we only consider those members where $\beta < \gamma$. The other symmetry where you interchange the first and second indices and also the third and fourth indices maps members of class G onto class H, and therefore none of the members of class H will be independent of class G. The symmetry $R_{\alpha[\beta\gamma\delta]} = 0$ implies that $R_{\alpha\beta\alpha\gamma} + R_{\alpha\gamma\beta\alpha} + R_{\alpha\alpha\gamma\beta} = 0$, but $R_{\alpha\alpha\gamma\beta} = 0$ because it is a member of class A. Therefore this symmetry implies $R_{\alpha\beta\alpha\gamma} = -R_{\alpha\gamma\beta\alpha}$. This associates a member of class I with each member of class G, and hence members of class I are not independent of members of class G either. This uses up all the symmetries that might reduce the number of independent members of class G, so we are left with 12 independent coefficients, which we take to be $\mathbf{R_{0102}}$, $\mathbf{R_{0103}}$, $\mathbf{R_{0203}}$, $\mathbf{R_{1012}}$, $\mathbf{R_{1013}}$, $\mathbf{R_{1213}}$, $\mathbf{R_{2021}}$, $\mathbf{R_{2023}}$, $\mathbf{R_{2123}}$, $\mathbf{R_{3031}}$, $\mathbf{R_{3032}}$, $\mathbf{R_{3132}}$.

Class H

As discussed in Section 5.4.2, the members of class H are all equal to a member of class G by symmetry under the interchange of the first and second indices as well as the interchange of the third and fourth indices. Hence there are no further independent Riemann coefficients in class H.

Class I

As discussed in Section 5.4.2, the members of class I are each the negative of a member of class H through the symmetry $R_{\alpha[\beta\gamma\delta]} = 0$. Hence there are no further independent Riemann coefficients in class I.

Class J

If you interchange the first and second indices only on a member of class H, then you get a member of class J. Hence there are no further independent Riemann coefficients in class J.

5.4.3 Independent Riemann Coefficients – Summary

Overall there are 20 independent coefficients of the Riemann tensor, which we take to be $\mathbf{R_{0101}}$, $\mathbf{R_{0202}}$, $\mathbf{R_{0303}}$, $\mathbf{R_{1212}}$, $\mathbf{R_{1313}}$, $\mathbf{R_{2323}}$, $\mathbf{R_{0123}}$, $\mathbf{R_{0213}}$, $\mathbf{R_{0102}}$, $\mathbf{R_{0103}}$, $\mathbf{R_{0203}}$, $\mathbf{R_{1012}}$, $\mathbf{R_{1013}}$, $\mathbf{R_{1213}}$, $\mathbf{R_{2021}}$, $\mathbf{R_{2023}}$, $\mathbf{R_{2123}}$, $\mathbf{R_{3031}}$, $\mathbf{R_{3032}}$, and $\mathbf{R_{3132}}$. Of the rest, 112 are zero, and the remaining 124 are related to these coefficients by symmetries. We emphasise that these results are quite general and apply to any four-dimensional space-time, not just the Schwarzschild solution. There are only ever a maximum of 20 independent components of the Riemann curvature tensor that must be evaluated.

5.5 The Ricci Curvature Tensor

The Riemann curvature has four free indices, and the stress energy tensor that we studied in Chapter 3 has only two. Can we make a contraction of the Riemann curvature so that the number of indices matches up? There are several candidates that we could try. For example, we could try raising the α index and contracting to form

$$g^{\beta\alpha} R_{\phi\lambda\alpha\beta} = R_{\phi\lambda}{}^{\beta}{}_{\beta}. \tag{5.25}$$

However, we know that the Riemann curvature is antisymmetric with respect to the interchange of the last two indices, so this contraction is zero. Similarly, if you raise the ϕ or λ indices and contract with the other of the first two indices, then that will give zero because of the antisymmetry of Equation (5.21).

More interesting is to raise the first index and then contract with the third one. It turns out that this is equivalent to other options such as raising the second index and contracting with the third, or raising the second index and contracting with the fourth, because of the symmetries from Equations (5.22) and (5.23). The Ricci curvature tensor is defined as

$$R_{\lambda\beta} = g^{\alpha\phi} R_{\phi\lambda\alpha\beta}$$
$$= R^{\alpha}{}_{\lambda\alpha\beta}. \tag{5.26}$$

Note that because of the symmetry property of the Riemann tensor from Equation (5.22), $R_{\alpha\beta\phi\lambda} = R_{\phi\lambda\alpha\beta}$, the Ricci tensor is symmetric:

$$R_{\beta\lambda} = R_{\lambda\beta}. \tag{5.27}$$

From the Ricci tensor we can also perform one more contraction by raising one of the remaining two indices and contracting:

$$R = g^{\beta\lambda} R_{\lambda\beta}. \tag{5.28}$$

This object is a tensor of rank $^0_0$ and is called the Ricci scalar. The Riemann curvature and the Ricci tensor and scalar are all the objects we will need to consider for describing the curvature of a space.

5.6 Bianchi Identities

The symmetry of the Riemann tensor shown in Equation (5.24) was derived by cyclically permuting the last three indices, adding up the results, and then showing that the resulting 12 terms cancelled, and therefore the sum of these three cyclically permuted Riemann tensor components is zero.

A very similar procedure can be used to prove the following, known as the Bianchi identities. Rather than repeating an almost identical proof, this one is left as an exercise for the reader in Problem 5.2. First, working in a flat space, you can prove that

$$\partial_\mu R_{\phi\lambda\alpha\beta} + \partial_\alpha R_{\phi\lambda\beta\mu} + \partial_\beta R_{\phi\lambda\mu\alpha} = 0. \tag{5.29}$$

Next, you re-write this as a tensor relationship by replacing ordinary derivatives with covariant ones. The result, valid in any coordinate system, is

$$D_\mu R_{\phi\lambda\alpha\beta} + D_\alpha R_{\phi\lambda\beta\mu} + D_\beta R_{\phi\lambda\mu\alpha} = 0. \tag{5.30}$$

These identities will be crucial to deriving Einstein's equations.

5.7 Einstein's Equations, up to a Constant

We now proceed to derive Einstein's equations. We recall from Chapter 3 that in a flat space the stress energy tensor for non-relativistic matter, $T^{\mu\nu}$, a tensor of rank $_0^2$, has zero divergence:

$$\frac{\partial T^{\mu\nu}}{\partial x^\nu} = 0. \tag{5.31}$$

In a curved space-time, we now replace the ordinary derivative with a covariant one, writing

$$D_\nu T^{\mu\nu} = 0. \tag{5.32}$$

Can we show that any combination of the tensors measuring curvature have the same properties? We apply the contraction used on the Riemann tensor in Equations (5.26)–(5.30):

$$g^{\alpha\phi}(D_\mu R_{\phi\lambda\alpha\beta} + D_\alpha R_{\phi\lambda\beta\mu} + D_\beta R_{\phi\lambda\mu\alpha}) = 0. \tag{5.33}$$

The $g^{\alpha\phi}$ raising operators commute with the covariant derivatives because the first derivatives of $g^{\alpha\phi}$ with respect to position vanish in a coordinate system tangent to a curved space – this is the same argument that led from Equation (5.12) to (5.14). Consequently, the first term becomes the covariant derivative of the Ricci tensor:

$$D_\mu R_{\lambda\beta} + D_\alpha g^{\alpha\phi} R_{\phi\lambda\beta\mu} + D_\beta g^{\alpha\phi} R_{\phi\lambda\mu\alpha} = 0. \tag{5.34}$$

For the last term, we recall the symmetry property of the Riemann tensor from Equation (5.13) and switch the final two indices, resulting in a sign flip:

$$D_\mu R_{\lambda\beta} + D_\alpha g^{\alpha\phi} R_{\phi\lambda\beta\mu} - D_\beta g^{\alpha\phi} R_{\phi\lambda\alpha\mu} = 0. \tag{5.35}$$

Applying the raising operators, we arrive at

$$D_\mu R_{\lambda\beta} + D_\alpha R^\alpha{}_{\lambda\beta\mu} - D_\beta R_{\lambda\mu} = 0. \tag{5.36}$$

We apply the second raising operator $g^{\beta\lambda}$:

$$g^{\beta\lambda} D_\mu R_{\lambda\beta} + g^{\beta\lambda} D_\alpha R^\alpha{}_{\lambda\beta\mu} - g^{\beta\lambda} D_\beta R_{\lambda\mu} = 0. \tag{5.37}$$

As before, the raising operators commute with the covariant derivatives:

$$D_\mu g^{\beta\lambda} R_{\lambda\beta} + D_\alpha g^{\beta\lambda} R^\alpha{}_{\lambda\beta\mu} - D_\beta g^{\beta\lambda} R_{\lambda\mu} = 0$$
$$D_\mu R + D_\alpha R^{\alpha\beta}{}_{\beta\mu} - D_\beta R^\beta{}_\mu = 0. \tag{5.38}$$

We now interchange the first two indices in $R^{\alpha\beta}{}_{\beta\mu}$, which yields $-R^{\beta\alpha}{}_{\beta\mu} = -R^\alpha{}_\mu$:

$$D_\mu R - D_\alpha R^\alpha{}_\mu - D_\beta R^\beta{}_\mu = 0$$
$$D_\mu R - 2D_\alpha R^\alpha{}_\mu = 0. \tag{5.39}$$

Inserting a Kronecker delta, we can write this as

$$D_\alpha \delta^\alpha_\mu R - 2D_\alpha R^\alpha{}_\mu = 0. \qquad (5.40)$$

Writing $\delta^\alpha_\mu = g_{\mu\phi}g^{\alpha\phi}$ and $R^\alpha{}_\mu = g_{\mu\phi}R^{\alpha\phi}$ and factoring the common $g_{\mu\phi}$ out, we obtain

$$-2g_{\mu\phi}D_\alpha\left(R^{\alpha\phi} - \frac{g^{\alpha\phi}}{2}R\right) = 0. \qquad (5.41)$$

We define

$$G^{\alpha\phi} = R^{\alpha\phi} - \frac{g^{\alpha\phi}}{2}R. \qquad (5.42)$$

Note that since both $R^{\alpha\phi}$ and $g^{\alpha\phi}$ are symmetric with respect to interchange of their indices, so is $G^{\alpha\phi}$, and we can therefore write

$$D_\alpha G^{\alpha\phi} = D_\alpha G^{\phi\alpha} = 0. \qquad (5.43)$$

The components $G^{\alpha\phi}$ are those of the Einstein tensor. They are symmetric in the interchange of the indices and have zero divergence. The Einstein tensor contains information about the curvature of space-time. The components of the stress energy tensor have the same symmetry properties and rank and also zero divergence. This led Einstein to propose that the stress energy could be interpreted as the source for the curvature:

$$G^{\alpha\beta} = AT^{\alpha\beta}, \qquad (5.44)$$

where A is some proportionality constant, as yet unknown. These are Einstein's equations.

Einstein's equations are the mathematical expression of Wheeler's quotation from Chapter 1. On the left-hand side, there is a measure of the curvature of space-time, mathematically defined in terms of closed-loop parallel transport around loops aligned with the basis vectors in the chosen coordinate system. On the right-hand side, there is the tensor expression of energy, momentum, and stress in space-time, which we derived for dust, but for which other constituents of the universe will contribute stress energy tensors of the same rank and with the same symmetry properties.

We have therefore achieved one objective of the book, to put a mathematical structure alongside Wheeler's statement. We now know how to describe, mathematically, the bending of space-time and the matter and energy content of the Universe.

We still have to figure out the value of the constant A. Armed with a knowledge of the physical constants that are likely to feature in this constant, you can show if you do Problem 5.7 that A can be written as a dimensionless number times G/c^4.

However, we cannot go further using this method to work out the value of the dimensionless number. This step will be addressed as part of our study of cosmology in Chapter 7.

5.8 Problems

5.1 The tangent space to a line of constant colatitude $\theta = \theta_0$ on spherical surface of radius R is made by dropping a cone of the appropriate geometry onto the sphere such that its inside surface touches the sphere at all points on this colatitude line (the colatitude is ninety degrees minus the latitude, which is defined as the angle north or south of the equator instead of downwards from the north pole). The cone can be cut from its apex vertically down the cone, making a right angle with the line of constant colatitude, and the piece of paper flattened out. This leaves a sector of a circle with a missing segment of opening angle α. See Figure 5.2.

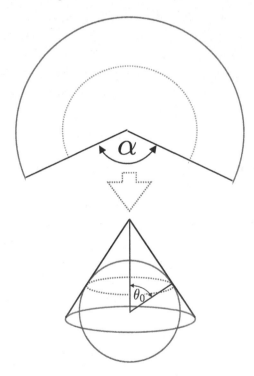

Figure 5.2 Tangent space to a line of constant latitude on a sphere.

(a) Calculate α in terms of θ_0 for any θ_0 in the range $[0, \pi/2)$.
(b) A vector $\vec{A}$ is parallel transported around the line of constant latitude, and at a point with coordinates (θ_0, ϕ) on this line has components A^θ and

A^ϕ. Using the Christoffel symbols in spherical polar coordinates derived in homework 1, derive expressions for $dA^\theta/d\phi$ and $dA^\phi/d\phi$ as functions of some or all of θ_0, ϕ, and R.

(c) Solve the equations from part (b) to obtain expressions for A^θ and A^ϕ as functions of the longitude ϕ. Predict the result for difference in the orientations of the vector after, compared to before, a round trip along the colatitude line, and show that the prediction for this result agrees with that you would expect from parallel transport on the flattened circular sector based purely on geometric arguments.

5.2 By analogy with the procedure for deriving Equation (5.24) prove the Bianchi identities of Equation (5.30).

5.3 Show that on the surface of a sphere, twelve out of the sixteen components of the Riemann curvature with all components lowered $R_{\phi\lambda\beta\alpha}$ are zero, and show how the remaining components are related to each other.

5.4 Using Equation (5.12), deduce the Riemann curvature component $R^\theta{}_{\phi\theta\phi}$, again on the surface of a sphere.

5.5 Find all the other non-zero Riemann curvature coefficients on the surface of the sphere with the first of the four indices raised. Do this without further explicit calculation using Equation (5.12). Instead, deduce them from the results in Question 5.3.

5.6 What are the dimensions of the Christoffel symbols in terms of some, or all, of the symbols M for mass, L for length, and T for time? From this deduce the dimensions of the components of the Einstein tensor $G^{\alpha\beta}$. What are the dimensions of the components of the stress energy tensor $T^{\alpha\beta}$ in terms of the same symbols? Hence deduce the dimensions of the constant A in Equation (5.44). Hence show that the dimensions of A are the same as those of the ratio of Newton's gravitational constant to the fourth power of the speed of light.

5.7 Show by applying dimensional analysis to Einstein's equations that the dimensions of the constant A are those of G/c^4.

6

Schwarzschild's Solution

6.1 Introduction

Karl Schwarzschild was in his early forties when he found the first exact solution to Einstein's equations. The problem he set out to solve was to identify the components of a metric that is spherically symmetric and time-independent. This must be an important problem, because the Universe is full of bodies with greater and lesser degrees of spherical symmetry, and whose mass distribution varies so slowly with time, that the gravitational field, and hence the metric tensor surrounding these bodies, must have these properties.

The task is conceptually simple but complex in practice. We first write down the most general metric that has the above properties. From this metric we deduce the Christoffel symbols, from which we compute the coefficients of the Riemann, Ricci, and Einstein tensors. The requirement that the Einstein tensor coefficients must satisfy the Einstein equations (5.44) constrains the form of the general metric coefficients. Finally, we require that the metric has the right limiting behaviour as $r \to \infty$ and yields the acceleration of a massive body predicted by Newton's law of gravitation for a slowly moving body. This brings us to Schwarzschild's solution.

6.2 General Form of the Metric

First, we give the general form of the metric. Imagining that the matter distribution giving rise to this metric will be close to the origin of coordinates and that far from this object, space-time will tend to being flat, we require that in the limit of large r, the metric signature of space-time in spherical polar coordinates becomes

$$ds^2 \to -c^2\,dt^2 + dr^2 + r^2\,d\theta^2 + r^2\sin^2\theta\,d\phi^2, \tag{6.1}$$

where (r, θ, ϕ) are the usual spherical polar coordinates. We attempt to preserve the sign of ds^2 at all values of r, since modifications to the coordinates should not change the sign of the terms in Equation (6.1). We will see presently that this attempt meets with only limited success.

All modifications should depend on r only, so that the resulting metric remains spherically symmetric. We avoid introducing off-diagonal terms that either mix the spatial and time coordinates or mix the different spatial coordinates – the latter because we want Pythagoras' theorem to remain true for right-angled triangles in the spatial coordinates. Finally, we ask that the r coordinate in the modified coordinate system under consideration preserve the area of a sphere centered on the origin at radius r as $4\pi r^2$.

The most general metric signature satisfying these requirements can be written

$$ds^2 = -e^{2\Phi(r)}c^2\,dt^2 + e^{2\Lambda(r)}\,dr^2 + r^2\,d\theta^2 + r^2\sin^2\theta\,d\phi^2. \tag{6.2}$$

The use of exponentials in the new terms $e^{2\Phi(r)}$ and $e^{2\Lambda(r)}$ constitutes our attempt to ensure that these terms are always positive, so that the signs of the g_{tt} and g_{rr} metric coefficients do not change with radius. Starting with this educated guess of a metric signature, let us calculate all derived non-tensor and tensor objects and see where the trail leads.

6.3 Christoffel Symbols

Our first task is to evaluate the Christoffel symbols. The Lagrangian is

$$L = \left[e^{2\Phi(r)}\left(\frac{d\,ct}{d\tau}\right)^2 - e^{2\Lambda(r)}\left(\frac{dr}{d\tau}\right)^2 - r^2\left(\frac{d\theta}{d\tau}\right)^2 - r^2\sin^2\theta\left(\frac{d\phi}{d\tau}\right)^2 \right]^{1/2}. \tag{6.3}$$

We take the partial derivative of L with respect to $d(ct)/d\tau$,

$$\frac{\partial L}{\partial(d\,ct/d\tau)} = \frac{e^{2\Phi}}{L}\frac{d(ct)}{d\tau}, \tag{6.4}$$

and differentiate again with respect to τ:

$$\frac{d}{d\tau}\frac{\partial L}{\partial(d\,ct/d\tau)} = \frac{e^{2\Phi}}{L}2\frac{d\Phi}{d\tau}\frac{d(ct)}{d\tau} + \frac{e^{2\Phi}}{L}\frac{d^2(ct)}{d\tau^2}$$

$$= \frac{e^{2\Phi}}{L}\left(\frac{d^2(ct)}{d\tau^2} + 2\frac{d\Phi}{dr}\frac{dr}{d\tau}\frac{d(ct)}{d\tau}\right), \tag{6.5}$$

where in the final step, we have rewritten $d\Phi/d\tau$ using the chain rule. There is no explicit dependence of L on the t coordinate, so $\partial L/\partial t = 0$, and the Euler–Lagrange equation in the time coordinate can be written

$$\frac{d^2(ct)}{d\tau^2} + 2\frac{d\Phi}{dr}\frac{dr}{d\tau}\frac{d(ct)}{d\tau} = 0. \tag{6.6}$$

Comparing with Equations (4.86), we read off our first two non-zero Christoffel symbols. We use a labelling system where numbers are used for the coordinates, as this is less confusing when we come to sum over indices, so that $0 = t$, $1 = r$, $2 = \theta$, and $3 = \phi$. With this convention,

$$\Gamma^0_{10} = \Gamma^0_{01} = \frac{d\Phi}{dr}. \tag{6.7}$$

All other Christoffel symbols with an upstairs 0 are zero.

Moving on to the r, or 1-component, we partially differentiate the Lagrangian with respect to $dr/d\tau$, obtaining

$$\frac{\partial L}{\partial(dr/d\tau)} = -\frac{e^{2\Lambda}}{L}\frac{dr}{d\tau}, \tag{6.8}$$

and then take the ordinary derivative with respect to τ, yielding

$$\begin{aligned}
\frac{d}{d\tau}\frac{\partial L}{\partial(dr/d\tau)} &= -\frac{1}{L}\left(2\frac{d\Lambda}{d\tau}e^{2\Lambda}\frac{dr}{d\tau} + e^{2\Lambda}\frac{d^2r}{d\tau^2}\right) \\
&= -\frac{e^{2\Lambda}}{L}\left(\frac{d^2r}{d\tau^2} + 2\frac{d\Lambda}{dr}\frac{dr}{d\tau}\frac{dr}{d\tau}\right).
\end{aligned} \tag{6.9}$$

We next evaluate $\partial L/\partial r$, obtaining

$$\begin{aligned}
\frac{\partial L}{\partial r} = \frac{1}{2L}&\left(2\frac{d\Phi}{dr}e^{2\Phi}\left(\frac{d(ct)}{d\tau}\right)^2 - 2\frac{d\Lambda}{dr}e^{2\Lambda}\left(\frac{dr}{d\tau}\right)^2\right. \\
&\left. - 2r\left(\frac{d\theta}{d\tau}\right)^2 - 2r\sin^2\theta\left(\frac{d\phi}{d\tau}\right)^2\right).
\end{aligned} \tag{6.10}$$

Equations (6.9) and (6.10) lead to the second Euler–Lagrange equation

$$\begin{aligned}
\frac{d}{d\tau}\frac{\partial L}{\partial(dr/d\tau)} - \frac{\partial L}{\partial r} = -\frac{e^{2\Lambda}}{L}&\left(\frac{d^2r}{d\tau^2} + \frac{d\Lambda}{dr}\left(\frac{dr}{d\tau}\right)^2 + \frac{d\Phi}{dr}e^{2(\Phi-\Lambda)}\left(\frac{d(ct)}{d\tau}\right)^2\right. \\
&\left. - re^{-2\Lambda}\left(\frac{d\theta}{d\tau}\right)^2 - re^{-2\Lambda}\sin^2\theta\left(\frac{d\phi}{d\tau}\right)^2\right) = 0, \tag{6.11}
\end{aligned}$$

from which, again by comparison with Equation (4.86), we read off four more non-zero Christoffel symbols:

$$\begin{aligned}
\Gamma^1_{00} &= e^{2(\Phi-\Lambda)}\frac{d\Phi}{dr} & \Gamma^1_{11} &= \frac{d\Lambda}{dr} \\
\Gamma^1_{22} &= -re^{-2\Lambda} & \Gamma^1_{33} &= -r\sin^2\theta\, e^{-2\Lambda}. \tag{6.12}
\end{aligned}$$

We turn next to the θ coordinate. Here we have

$$\frac{\partial L}{\partial(d\theta/d\tau)} = -\frac{1}{2L}2r^2\frac{d\theta}{d\tau} - \frac{-r^2}{L}\frac{d\theta}{d\tau}, \tag{6.13}$$

and its time derivative is

$$\frac{d}{d\tau}\frac{\partial L}{\partial(d\theta/d\tau)} = -\frac{1}{L}\left(2r\frac{dr}{d\tau}\frac{d\theta}{d\tau} + r^2\frac{d^2\theta}{d\tau^2}\right)$$

$$= -\frac{r^2}{L}\left(\frac{d^2\theta}{d\tau^2} + \frac{2}{r}\frac{dr}{d\tau}\frac{d\theta}{d\tau}\right). \tag{6.14}$$

The partial derivative of L with respect to θ is

$$\frac{\partial L}{\partial \theta} = -\frac{1}{2L}\left(2r^2\sin\theta\cos\theta\left(\frac{d\phi}{d\tau}\right)^2\right), \tag{6.15}$$

and hence the Euler–Lagrange equation for the θ coordinate is

$$\frac{d^2\theta}{d\tau^2} + \frac{2}{r}\frac{dr}{d\tau}\frac{d\theta}{d\tau} - \sin\theta\cos\theta\left(\frac{d\phi}{d\tau}\right)^2 = 0. \tag{6.16}$$

Again comparing with the geodesic equations, we read off three more non-zero Christoffel symbols:

$$\Gamma^2_{12} = \Gamma^2_{21} = \frac{1}{r} \qquad\qquad \Gamma^2_{33} = -\sin\theta\cos\theta. \tag{6.17}$$

Finally, the ϕ coordinate:

$$\frac{\partial L}{\partial(d\phi/d\tau)} = \frac{1}{2L}\left(-2r^2\sin^2\theta\frac{d\phi}{d\tau}\right) = -\frac{r^2\sin^2\theta}{L}\left(\frac{d\phi}{d\tau}\right), \tag{6.18}$$

so that

$$\frac{d}{d\tau}\frac{\partial L}{\partial(d\phi/d\tau)}$$

$$= -\frac{1}{L}\left(2r\frac{dr}{d\tau}\frac{d\phi}{d\tau}\sin^2\theta + 2r^2\sin\theta\cos\theta\frac{d\theta}{d\tau}\frac{d\phi}{d\tau} + r^2\sin^2\theta\frac{d^2\phi}{d\tau^2}\right), \tag{6.19}$$

and ϕ is a cyclic coordinate, so $\partial L/\partial\phi = 0$. Therefore the Euler–Lagrange equation in ϕ is

$$\frac{d^2\phi}{d\tau^2} + \frac{2}{r}\frac{dr}{d\tau}\frac{d\phi}{d\tau} + \frac{2}{\tan\theta}\frac{d\theta}{d\tau}\frac{d\phi}{d\tau} = 0, \tag{6.20}$$

so that the final four non-zero Christoffel symbols are

$$\Gamma^3_{13} = \Gamma^3_{31} = \frac{1}{r} \qquad\qquad \Gamma^3_{23} = \Gamma^3_{32} = \frac{1}{\tan\theta}. \tag{6.21}$$

Gathering all of the non-zero Christoffel symbols together in one space for convenience, we have

$$\Gamma^0_{10} = \Gamma^0_{01} = \frac{d\Phi}{dr}$$

$$\Gamma^1_{00} = e^{2(\Phi-\Lambda)}\frac{d\Phi}{dr} \qquad\qquad \Gamma^1_{11} = \frac{d\Lambda}{dr}$$

$$\Gamma^1_{22} = -re^{-2\Lambda} \qquad\qquad \Gamma^1_{33} = -r\sin^2\theta\, e^{-2\Lambda}$$

$$\Gamma^2_{12} = \Gamma^2_{21} = \frac{1}{r} \qquad\qquad \Gamma^2_{33} = -\sin\theta\cos\theta$$

$$\Gamma^3_{13} = \Gamma^3_{31} = \frac{1}{r} \qquad\qquad \Gamma^3_{23} = \Gamma^3_{32} = \frac{1}{\tan\theta}. \qquad (6.22)$$

6.4 Evaluation of Riemann Components

In Chapter 5, we introduced the Riemann curvature and derived Equation (5.12), an expression for the components in terms of Christoffel symbols and their spatial derivatives. This relatively complicated expression was then simplified by assuming that our coordinate system of choice was tangent to a geodesic, which I refer to as pigeon coordinates, so that the first derivatives of the metric coefficients, and hence the terms containing only Christoffel symbols and not their derivatives, are zero. This eliminated all but two of the terms on the right-hand side of Equation (5.12). Finally, the one upstairs index was lowered leading to Equation (5.19).

From this latter equation we extracted five symmetries between the components $R_{\phi\lambda\beta\alpha}$, making it clear that many of the 256 components of the Riemann curvature are in fact zero and also that amongst the rest, a very much smaller number of components, 20 in fact, are independent of each other. The conclusion is that knowledge of the 20 Riemann coefficients $\mathbf{R_{0102}}$, $\mathbf{R_{0103}}$, $\mathbf{R_{0203}}$, $\mathbf{R_{1012}}$, $\mathbf{R_{1013}}$, $\mathbf{R_{1213}}$, $\mathbf{R_{2021}}$, $\mathbf{R_{2023}}$, $\mathbf{R_{2123}}$, $\mathbf{R_{3031}}$, $\mathbf{R_{3032}}$, and $\mathbf{R_{3132}}$ is sufficient to deduce the values of all the other non-zero coefficients. We refer to these from now on as the independent coefficients. We could also choose many other sets of 20 independent coefficients, but we will use this one set in both this chapter and the next.

6.4.1 Evaluating Independent Components

In one sense, the treatment between Equations (5.12) and (5.19) is over-simplistic. In reality, we are rarely in a coordinate system tangent to the world line of a particle. The tangent space of 'pigeon' coordinates was useful for finding the symmetries amongst the Riemann components, and these symmetries hold more generally in any coordinate system because the statements of the symmetries with respect to index interchanges, and also the Bianchi identities, are tensor statements. However, in the coordinate systems that are used in practice, we will not in general be in a 'pigeon' frame of reference! We will need to evaluate the 20 independent Riemann components using Equation (5.12), an equation that has ten terms, and then relate them to other components using the symmetries and raising and lowering indices using the metric tensor coefficients. Once we have the components of the Riemann tensor, we can use them to deduce the components of the Ricci tensor, the Ricci scalar, and finally the Einstein tensor.

This task is made much simpler by the fact that many metrics of interest have all of-diagonal terms zero, and it will turn out that most of the remaining independent Riemann tensor coefficients are zero. However, as we will see, even in these cases, the evaluation of the tensor components is lengthy and arduous. The author believes that nonetheless students should go through the calculation in detail for at least one or two cases. Once they appreciate the mathematical steps, they can if they wish employ computer codes that do the mathematical steps automatically. Some insights that may be missed in stepping over the analytical work may be compensated for by the time saved and the mistakes that will be avoided by having a computer do the work for you. Computer codes can also be buggy and give incorrect answers. Perhaps, most valuable is to use both a computer code and manual algebra, checking one method with the other.

We reproduce here Equation (5.12) for convenience:

$$R^{\mu}{}_{\lambda\beta\alpha} = \frac{\partial \Gamma^{\mu}_{\alpha\lambda}}{\partial x^{\beta}} - \frac{\partial \Gamma^{\mu}_{\beta\lambda}}{\partial x^{\alpha}} + \Gamma^{\mu}_{\beta\kappa}\Gamma^{\kappa}_{\alpha\lambda} - \Gamma^{\mu}_{\alpha\kappa}\Gamma^{\kappa}_{\beta\lambda}. \qquad (6.23)$$

We also reproduce the Christoffel symbols for the hypothesised metric from Equation (6.22):

$$\Gamma^0_{10} = \Gamma^0_{01} = \frac{d\Phi}{dr}$$

$$\Gamma^1_{00} = e^{2(\Phi-\Lambda)}\frac{d\Phi}{dr} \qquad\qquad \Gamma^1_{11} = \frac{d\Lambda}{dr}$$

$$\Gamma^1_{22} = -re^{-2\Lambda} \qquad\qquad \Gamma^1_{33} = -r\sin^2\theta\, e^{-2\Lambda}$$

$$\Gamma^2_{12} = \Gamma^2_{21} = \frac{1}{r} \qquad\qquad \Gamma^2_{33} = -\sin\theta\cos\theta$$

$$\Gamma^3_{13} = \Gamma^3_{31} = \frac{1}{r} \qquad\qquad \Gamma^3_{23} = \Gamma^3_{32} = \frac{1}{\tan\theta}. \qquad (6.24)$$

We start with $R^0{}_{101}$, which from Equation (5.12) can be calculated from the Christoffel symbols of Equation (6.22) as follows:

$$R^0{}_{101} = \frac{\partial \Gamma^0_{11}}{\partial x^0} - \frac{\partial \Gamma^0_{01}}{\partial x^1} + \Gamma^0_{00}\Gamma^0_{11} + \Gamma^0_{01}\Gamma^1_{11} + \Gamma^0_{12}\Gamma^2_{11} + \Gamma^0_{13}\Gamma^3_{11}$$
$$- \Gamma^0_{10}\Gamma^0_{01} - \Gamma^0_{11}\Gamma^1_{01} - \Gamma^0_{12}\Gamma^2_{01} - \Gamma^0_{13}\Gamma^3_{01}. \qquad (6.25)$$

All the terms vanish except three, so that we are left with $-\partial\Gamma^0_{01}/\partial r + \Gamma^0_{01}\Gamma^1_{11} - \Gamma^0_{10}\Gamma^0_{01}$. Substituting in the Christoffel symbols, we obtain

$$R^0{}_{101} = -\frac{d^2\Phi}{dr^2} + \frac{d\Phi}{dr}\frac{d\Lambda}{dr} - \left(\frac{d\Phi}{dr}\right)^2. \tag{6.26}$$

For $R^0{}_{202}$, we get

$$R^0{}_{202} = \frac{\partial\Gamma^0_{22}}{\partial x^0} - \frac{\partial\Gamma^0_{02}}{\partial x^2} + \Gamma^0_{00}\Gamma^0_{22} + \Gamma^0_{01}\Gamma^1_{22} + \Gamma^0_{02}\Gamma^2_{22} + \Gamma^0_{03}\Gamma^3_{22}$$
$$- \Gamma^0_{20}\Gamma^0_{02} - \Gamma^0_{21}\Gamma^1_{02} - \Gamma^0_{22}\Gamma^2_{02} - \Gamma^0_{23}\Gamma^3_{02}. \tag{6.27}$$

All the terms vanish except $\Gamma^0_{01}\Gamma^1_{22}$. Substituting in the Christoffel symbols, we obtain

$$R^0{}_{202} = -re^{-2\Lambda}\frac{d\Phi}{dr}. \tag{6.28}$$

For $R^0{}_{303}$, we get

$$R^0{}_{303} = \frac{\partial\Gamma^0_{33}}{\partial x^0} - \frac{\partial\Gamma^0_{03}}{\partial x^3} + \Gamma^0_{00}\Gamma^0_{33} + \Gamma^0_{01}\Gamma^1_{33} + \Gamma^0_{02}\Gamma^2_{33} + \Gamma^0_{03}\Gamma^3_{33}$$
$$- \Gamma^0_{30}\Gamma^0_{03} - \Gamma^0_{31}\Gamma^1_{03} - \Gamma^0_{32}\Gamma^2_{03} - \Gamma^0_{33}\Gamma^3_{03}. \tag{6.29}$$

All the terms vanish except $\Gamma^0_{01}\Gamma^1_{33}$. Substituting in the Christoffel symbols, we obtain

$$R^0{}_{303} = -re^{-2\Lambda}\frac{d\Phi}{dr}\sin^2\theta. \tag{6.30}$$

For $R^1{}_{212}$, we get

$$R^1{}_{212} = \frac{\partial\Gamma^1_{22}}{\partial x^1} - \frac{\partial\Gamma^1_{21}}{\partial x^2} + \Gamma^1_{10}\Gamma^0_{22} + \Gamma^1_{11}\Gamma^1_{22} + \Gamma^1_{12}\Gamma^2_{22} + \Gamma^1_{13}\Gamma^3_{22}$$
$$- \Gamma^1_{20}\Gamma^0_{12} - \Gamma^1_{21}\Gamma^1_{12} - \Gamma^1_{22}\Gamma^2_{12} - \Gamma^1_{23}\Gamma^3_{12}. \tag{6.31}$$

All the terms vanish except $\partial\Gamma^1_{22}/\partial x^1 + \Gamma^1_{11}\Gamma^1_{22} - \Gamma^1_{22}\Gamma^2_{12}$. Substituting in the Christoffel symbols, we obtain

$$R^1{}_{212} = \frac{d}{dr}\left(-re^{-2\Lambda}\right) - r\frac{d\Lambda}{dr}e^{-2\Lambda} + re^{-2\Lambda}\frac{1}{r}$$
$$= r\frac{d\Lambda}{dr}e^{-2\Lambda}. \tag{6.32}$$

For $R^1{}_{313}$, we get

$$R^1{}_{313} = \frac{\partial\Gamma^1_{33}}{\partial x^1} - \frac{\partial\Gamma^1_{31}}{\partial x^3} + \Gamma^1_{10}\Gamma^0_{33} + \Gamma^1_{11}\Gamma^1_{33} + \Gamma^1_{12}\Gamma^2_{33} + \Gamma^1_{13}\Gamma^3_{33}$$
$$- \Gamma^1_{30}\Gamma^0_{13} - \Gamma^1_{31}\Gamma^1_{13} - \Gamma^1_{32}\Gamma^2_{13} - \Gamma^1_{33}\Gamma^3_{13}. \tag{6.33}$$

All terms vanish except $\partial\Gamma^1_{33}/\partial x^1 + \Gamma^1_{11}\Gamma^1_{33} - \Gamma^1_{33}\Gamma^3_{13}$. Substituting in the Christoffel symbols, we obtain

$$R^1{}_{313} = \frac{\partial}{\partial r}\left(-re^{-2\Lambda}\sin^2\theta\right) + \frac{d\Lambda}{dr}\left(-re^{-2\Lambda}\sin^2\theta\right) - \left(-re^{-2\Lambda}\sin^2\theta\,\frac{1}{r}\right)$$

$$= r\frac{d\Lambda}{dr}e^{-2\Lambda}\sin^2\theta. \tag{6.34}$$

For $R^2{}_{323}$, we get

$$R^2{}_{323} = \frac{\partial\Gamma^2_{33}}{\partial x^2} - \frac{\partial\Gamma^2_{23}}{\partial x^3} + \Gamma^2_{20}\Gamma^0_{33} + \Gamma^2_{21}\Gamma^1_{33} + \Gamma^2_{22}\Gamma^2_{33} + \Gamma^2_{23}\Gamma^3_{33}$$

$$- \Gamma^2_{30}\Gamma^0_{23} - \Gamma^2_{31}\Gamma^1_{23} - \Gamma^2_{32}\Gamma^2_{23} - \Gamma^2_{33}\Gamma^3_{23}. \tag{6.35}$$

All terms vanish except $\partial\Gamma^2_{33}/\partial x^2 + \Gamma^2_{21}\Gamma^1_{33} - \Gamma^2_{33}\Gamma^3_{23}$. Substituting in the Christoffel symbols, we obtain

$$R^2{}_{323} = \frac{\partial}{\partial\theta}\left(-\sin\theta\cos\theta\right) + \frac{1}{r}\left(re^{-2\Lambda}\sin^2\theta\right) - \left(-\sin\theta\cos\theta\right)\frac{\cos\theta}{\sin\theta}$$

$$= \left(1 - e^{-2\Lambda}\right)\sin^2\theta. \tag{6.36}$$

Following the same procedure for the remaining independent Riemann curvature coefficients, R_{0123}, R_{0213}, R_{0102}, R_{0103}, R_{0203}, R_{1012}, R_{1013}, R_{1213}, R_{2021}, R_{2023}, R_{2123}, R_{3031}, R_{3032}, and R_{3132}, we can show that all the terms in all these coefficients vanish either because at least one of the two Christoffel symbols in the term is zero or because a particular partial derivative of a Christoffel symbol is zero.

6.5 Ricci Tensor Coefficients

The Ricci coefficients are contractions across the Riemann coefficients. There are fortunately only ten independent ones. There is nothing for it but to go through and calculate them all. Let us start then at R_{00}. As before, we will do the first few very carefully, then go through the rest more rapidly, leaving it to the diligent student to go over the details carefully as a way of practicing the algebra. We have

$$R_{00} = R^0{}_{000} + R^1{}_{010} + R^2{}_{020} + R^3{}_{030}, \tag{6.37}$$

where $R^0{}_{000} = 0$ because all the indices are equal, placing its downstairs-index counterpart in class C. The other three are non-zero and related to the components we calculated above, in this case, $R^0{}_{101}$. However, we need to convert from $R^0{}_{101}$ to $R^1{}_{010}$. We do this by raising and lowering with the components of the metric. However, recall that the metric components are those given by Equation (6.2) reproduced here:

$$ds^2 = -e^{2\Phi(r)}c^2\,dt^2 + e^{2\Lambda(r)}\,dr^2 + r^2\,d\theta^2 + r^2\sin^2\theta\,d\phi^2, \tag{6.38}$$

so that only the diagonal components of $g_{\mu\nu}$ are non-zero and, along with their inverses, are

$$
\begin{aligned}
g_{00} &= -e^{2\Phi} & g^{00} &= -e^{-2\Phi} \\
g_{11} &= e^{2\Lambda} & g^{11} &= e^{-2\Lambda} \\
g_{22} &= r^2 & g^{22} &= \frac{1}{r^2} \\
g_{33} &= r^2 \sin^2\theta & g^{33} &= \frac{1}{r^2 \sin^2\theta}.
\end{aligned}
\tag{6.39}
$$

So we proceed with $R^1{}_{010}$:

$$
\begin{aligned}
R^0{}_{101} &= g^{00} R_{0101} \\
&= g^{00} R_{1010} \\
&= g^{00} g_{11} R^1{}_{010},
\end{aligned}
\tag{6.40}
$$

where we have used the symmetry $R_{\beta\alpha\delta\gamma} = R_{\alpha\beta\gamma\delta}$ between the first and second lines. Notice how much harder this would be if there were off-diagonal non-zero metric coefficients! Then each raising and lowering operation would also require a linear superposition. Consequently, we can write

$$
\begin{aligned}
R^1{}_{010} &= \frac{R^0{}_{101}}{g^{00} g_{11}} = g_{00} g^{11} R^0{}_{101} \\
&= e^{2(\Phi-\Lambda)} \left(\frac{d^2\Phi}{dr^2} - \frac{d\Phi}{dr}\frac{d\Lambda}{dr} + \left(\frac{d\Phi}{dr}\right)^2 \right).
\end{aligned}
\tag{6.41}
$$

We proceed similarly for $R^2{}_{020}$:

$$
\begin{aligned}
R^2{}_{020} &= g^{22} g_{00} R^0{}_{202} \\
&= \frac{1}{r} e^{2(\Phi-\Lambda)} \frac{d\Phi}{dr},
\end{aligned}
\tag{6.42}
$$

and for $R^3{}_{030}$,

$$
\begin{aligned}
R^3{}_{030} &= g^{33} g_{00} R^0{}_{303} \\
&= \frac{1}{r} e^{2(\Phi-\Lambda)} \frac{d\Phi}{dr}.
\end{aligned}
\tag{6.43}
$$

Adding up the results of Equations (6.41), (6.42), and (6.43), we obtain

$$
R_{00} = e^{2(\Phi-\Lambda)} \left(\left(\frac{d\Phi}{dr}\right)^2 + \frac{d^2\Phi}{dr^2} - \frac{d\Phi}{dr}\frac{d\Lambda}{dr} + \frac{2}{r}\frac{d\Phi}{dr} \right).
\tag{6.44}
$$

We perform the same types of calculations to evaluate the off-diagonal Ricci components R_{01}, R_{02}, and R_{03}:

$$R_{01} = R^0{}_{001} + R^1{}_{011} + R^2{}_{021} + R^3{}_{031} = 0$$
$$R_{02} = R^0{}_{002} + R^1{}_{012} + R^2{}_{022} + R^3{}_{032} = 0$$
$$R_{03} = R^0{}_{003} + R^1{}_{013} + R^2{}_{023} + R^3{}_{033} = 0, \tag{6.45}$$

where in each case, all of the contributing Riemann components were previously found to be zero. We carry on to evaluate R_{11}:

$$R_{11} = R^0{}_{101} + R^1{}_{111} + R^2{}_{121} + R^3{}_{131}, \tag{6.46}$$

where $R^1{}_{111} = 0$, the other three are non-zero, and $R^0{}_{101}$ was already calculated. The other two can be related by symmetries and by raising and lowering to $R^1{}_{212}$ and $R^1{}_{313}$. The result is

$$R_{11} = -\left(\frac{d\Phi}{dr}\right)^2 - \frac{d^2\Phi}{dr^2} + \frac{d\Phi}{dr}\frac{d\Lambda}{dr} + \frac{2}{r}\frac{d\Lambda}{dr}. \tag{6.47}$$

The next two off-diagonal components R_{21} and R_{31} turn out to be zero because all the contributing Riemann coefficients are zero, following the same procedure as in Equation (6.45). We therefore move on to

$$R_{22} = R^0{}_{202} + R^2{}_{212} + R^2{}_{222} + R^3{}_{232}, \tag{6.48}$$

where $R^2{}_{222} = 0$, and the other three contributing Riemann coefficients are calculated as before, yielding

$$R_{22} = e^{-2\Lambda}\left(r\frac{d\Lambda}{dr} - r\frac{d\Phi}{dr} - 1\right) + 1. \tag{6.49}$$

Component R_{23} is zero by the same procedure as for all the other off-diagonal components. Finally, we move to

$$R_{33} = R^0{}_{303} + R^2{}_{313} + R^2{}_{323} + R^3{}_{333}, \tag{6.50}$$

where $R^3{}_{333} = 0$, and the other three components are calculated from the six non-zero Riemann components as before, yielding the result

$$R_{33} = r\sin^2\theta\, e^{-2\Lambda}\left(\frac{d\Lambda}{dr} - \frac{d\Phi}{dr}\right) + \left(1 - e^{-2\Lambda}\right)\sin^2\theta. \tag{6.51}$$

6.6 The Ricci Scalar

The Ricci scalar is

$$g^{\alpha\beta} R_{\alpha\beta} = g^{00} R_{00} + g^{11} R_{11} + g^{22} R_{22} + g^{33} R_{33}$$

$$= -e^{-2\Phi} e^{2(\Phi-\Lambda)} \left(\left(\frac{d\Phi}{dr}\right)^2 + \frac{d^2\Phi}{dr^2} - \frac{d\Phi}{dr}\frac{d\Lambda}{dr} + \frac{2\,d\alpha}{r\,dr} \right)$$

$$+ e^{-2\Lambda} \left(-\left(\frac{d\Phi}{dr}\right)^2 - \frac{d^2\Phi}{dr^2} + \frac{d\Phi}{dr}\frac{d\Lambda}{dr} + \frac{2\,d\Lambda}{r\,dr} \right)$$

$$+ \frac{e^{-2\Lambda}}{r^2} \left(r\frac{d\Lambda}{dr} - r\frac{d\Phi}{dr} - 1 \right) + \frac{1}{r^2}$$

$$+ \frac{1}{r^2 \sin^2\theta} \left(r \sin^2\theta \frac{e^{-2\Lambda}}{r} \left(\frac{d\Lambda}{dr} - \frac{d\Phi}{dr} \right) + \frac{\sin^2\theta}{r^2} \left(1 - e^{-2\Lambda} \right) \right)$$

$$R = -2e^{-2\Lambda} \left(\left(\frac{d\Phi}{dr}\right)^2 + \frac{d^2\Phi}{dr^2} - \frac{d\Phi}{dr}\frac{d\Lambda}{dr} + \frac{2}{r}\left(\frac{d\Phi}{dr} - \frac{d\Lambda}{dr} \right) + \frac{1}{r^2} \right)$$

$$+ \frac{2}{r^2}. \tag{6.52}$$

6.7 The Einstein Tensor

We now evaluate some of the components of the Einstein tensor $G_{\alpha\beta} = R_{\alpha\beta} - g_{\alpha\beta} R/2$. Starting with G_{00},

$$G_{00} = R_{00} - \frac{1}{2} g_{00} R$$

$$= e^{2(\Phi-\Lambda)} \left(\left(\frac{d\Phi}{dr}\right)^2 + \frac{d^2\Phi}{dr^2} - \frac{d\Phi}{dr}\frac{d\Lambda}{dr} + \frac{2}{r}\frac{d\Phi}{dr} \right)$$

$$+ \frac{e^{2\Phi}}{2} \left(-2e^{-2\Lambda} \left(\left(\frac{d\Phi}{dr}\right)^2 + \frac{d^2\Phi}{dr^2} - \frac{d\Phi}{dr}\frac{d\Lambda}{dr} + \frac{2}{r}\left(\frac{d\Phi}{dr} - \frac{d\Lambda}{dr} \right) + \frac{1}{r^2} \right) \right.$$

$$\left. + \frac{2}{r^2} \right)$$

$$= e^{2(\Phi-\Lambda)} \left(\frac{2\,d\Lambda}{r\,dr} - \frac{1}{r^2} \right) + \frac{e^{2\Phi}}{r^2}. \tag{6.53}$$

Outside the matter distribution, we seek solutions where the elements of the stress energy tensor $T_{\alpha\beta}$ are zero, so that we are solving the so-called vacuum Einstein's equations $G_{\alpha\beta} = 0$. Setting $G_{00} = 0$ implies that

$$e^{-2\Lambda} \left(\frac{2\,d\Lambda}{r\,dr} - \frac{1}{r^2} \right) + \frac{1}{r^2} = 0. \tag{6.54}$$

This is a differential equation for $\Lambda(r)$. Fortunately, it can be solved analytically. Multiply through by $e^{2\Lambda}$ and rearrange:

$$\frac{e^{-2\Lambda}}{r^2} - \frac{2}{r}\frac{d\Lambda}{dr}e^{-2\Lambda}e^{-2\Lambda} = \frac{1}{r^2}$$

$$e^{-2\Lambda} - 2r\frac{d\Lambda}{dr}e^{-2\Lambda} = 1$$

$$\frac{d}{dr}\left(re^{-2\Lambda}\right) = 1. \tag{6.55}$$

This equation can now be integrated:

$$re^{-2\Lambda} = r + R_0, \tag{6.56}$$

where R_0 is a constant of integration, and therefore

$$e^{-2\Lambda} = 1 + \frac{R_0}{r}. \tag{6.57}$$

We now consider G_{11}:

$$G_{11} = R_{11} - \frac{1}{2}g_{11}R$$

$$= -\left(\frac{d\Phi}{dr}\right)^2 - \frac{d^2\Phi}{dr^2} + \frac{d\Phi}{dr}\frac{d\Lambda}{dr} + \frac{2}{r}\frac{d\Lambda}{dr}$$

$$+ \frac{e^{2\Lambda}}{2}\left(2e^{-2\Lambda}\left(\left(\frac{d\Phi}{dr}\right)^2 + \frac{d^2\Phi}{dr^2} - \frac{d\Phi}{dr}\frac{d\Lambda}{dr} + \frac{2}{r}\left(\frac{d\Phi}{dr} - \frac{d\Lambda}{dr}\right)\right.\right.$$

$$\left.\left.+ \frac{1}{r^2}\right) - \frac{2}{r^2}\right)$$

$$G_{11} = \frac{2}{r}\frac{d\Phi}{dr} + \frac{1}{r^2} - \frac{e^{2\Lambda}}{r^2}. \tag{6.58}$$

Setting $G_{11} = 0$, we obtain

$$e^{-2\Lambda}\left(\frac{2}{r}\frac{d\Phi}{dr} + \frac{1}{r^2}\right) - \frac{1}{r^2} = 0. \tag{6.59}$$

This is an inhomogeneous differential equation as it contains both $\Phi(r)$ and $\Lambda(r)$, but compare it to Equation (6.54) reproduced here:

$$e^{-2\Lambda}\left(\frac{2}{r}\frac{d\Lambda}{dr} - \frac{1}{r^2}\right) + \frac{1}{r^2} = 0. \tag{6.60}$$

Adding these two equations together, there is a very large degree of cancellation, and we are left with

$$\frac{d\Phi}{dr} + \frac{d\Lambda}{dr} = 0. \tag{6.61}$$

This can again be integrated, leading to

$$\Lambda = -\Phi + B, \tag{6.62}$$

where B is an integration constant. Multiplying by two and taking the exponential of both sides, we get

$$e^{-2\Lambda} = e^{-2B}e^{2\Phi}. \tag{6.63}$$

Substituting in from Equation (6.57), we find that

$$e^{2\Phi} = e^{2B}\left(1 + \frac{R_0}{r}\right). \tag{6.64}$$

We are now ready to substitute Equations (6.57) and (6.64) back into Equation (6.2) to obtain

$$ds^2 = -c^2\,dt^2 e^{2B}\left(1 + \frac{R_0}{r}\right) + \frac{dr^2}{1 + R_0/r} + r^2\,d\theta^2 + r^2\sin^2\theta\,d\phi^2. \tag{6.65}$$

We require that at large r we recover the Minkowski metric. Therefore we must have $e^{2B} = 1$ or $B = 0$, so that

$$ds^2 = -c^2\,dt^2\left(1 + \frac{R_0}{r}\right) + \frac{dr^2}{1 + R_0/r} + r^2\,d\theta^2 + r^2\sin^2\theta\,d\phi^2. \tag{6.66}$$

This leaves one remaining integration constant R_0 to be evaluated.

6.8 Correspondence with Newtonian Gravity

We can work out the correspondence between the predictions of Schwarzschild's metric and Newtonian gravity for the case of a freely falling body on a radial trajectory. We start with the geodesic equation for the r (or 1) coordinate:

$$\frac{d^2x^1}{d\tau^2} + \Gamma^1_{\mu\nu}\frac{dx^\mu}{d\tau}\frac{dx^\nu}{d\tau} = 0. \tag{6.67}$$

We look up the non-zero Christoffel symbols in the Schwarzschild geometry from Equation (6.22). We see that the non-zero symbols with an upper 1 are Γ^1_{00}, Γ^1_{11}, Γ^1_{22}, and Γ^1_{22}. Therefore the geodesic equation for r becomes

$$\frac{d^2r}{d\tau^2} = -\Gamma^1_{00}\left(\frac{dx^0}{d\tau}\right)^2 - \Gamma^1_{11}\left(\frac{dx^1}{d\tau}\right)^2 - \Gamma^1_{22}\left(\frac{dx^2}{d\tau}\right)^2 - \Gamma^1_{33}\left(\frac{dx^3}{d\tau}\right)^2. \tag{6.68}$$

The last two terms correspond to angular motion because $x^2 = \theta$ and $x^3 = \phi$, so these terms do not contribute for a particle on a radial trajectory. We examine the other two terms, firstly the term in $x^0 = ct$. This term is proportional to Γ^1_{00}. We have

$$\Gamma^1_{00} = e^{2(\Phi - \Lambda)} \frac{d\Phi}{dr} = -e^{-2\Lambda} \frac{d\Lambda}{dr}, \tag{6.69}$$

where we have used $\Phi = -\Lambda$. We also know from Equation (6.57) that

$$e^{-2\Lambda} = 1 + \frac{R_0}{r}. \tag{6.70}$$

Differentiating this result with respect to r, we obtain

$$-2 \frac{d\Lambda}{dr} e^{-2\Lambda} = -\frac{R_0}{r^2}, \tag{6.71}$$

so that, comparing with Equation (6.69), we are able to express Γ^1_{00} in terms of A:

$$\Gamma^1_{00} = -\frac{R_0}{2r^2}. \tag{6.72}$$

The next Christoffel symbol is

$$\Gamma^1_{11} = \frac{d\Lambda}{dr}, \tag{6.73}$$

so that from Equations (6.70) and (6.71) we get

$$\Gamma^1_{11} = \frac{R_0/(2r^2)}{1 + R_0/r}. \tag{6.74}$$

Now, for a particle mass m in free fall in the Newtonian gravitational field of a body of mass M at the origin, the equation of motion is

$$\frac{-GMm}{r^2} = m \frac{d^2r}{dt^2}$$

$$\frac{d^2r}{dt^2} = -\frac{GM}{r^2}. \tag{6.75}$$

For a non-relativistic radial trajectory, the proper time τ is approximately equal to the coordinate time t. Substituting Equations (6.72), (6.74), and (6.75) into the geodesic equation (6.68), we obtain

$$\frac{GM}{r^2} = -\frac{R_0}{2r^2} \left(\frac{d(ct)}{dt} \right)^2 - \frac{R_0/(2r^2)}{1 + R_0/r} \left(\frac{dr}{dt} \right)^2, \tag{6.76}$$

or

$$\frac{2GM}{c^2} = -R_0 - \frac{R_0}{1 + R_0/r} \frac{(dr/dt)^2}{c^2}. \tag{6.77}$$

In this form, the final term is seen to be a relativistic correction, as it is proportional to the squared ratio of the radial component of the velocity of the falling body to the speed of light. We therefore neglect this term in the Newtonian limit and arrive at a value for the constant:

$$R_0 = \frac{-2GM}{c^2}. \tag{6.78}$$

We have now solved for all the unknown constants in the Schwarzschild metric. As we will see in Problem 6.1, we have not quite achieved all the goals for the solution laid out in Section 6.2, but everything is at least mathematically consistent. The final form of the metric signature for the Schwarzschild geometry is

$$ds^2 = -c^2 \, dt^2 \left(1 - \frac{2GM}{rc^2}\right) + \frac{dr^2}{1 - 2GM/(rc^2)}$$
$$+ r^2 \, d\theta^2 + r^2 \sin^2 \theta \, d\phi^2. \tag{6.79}$$

Often, we define the Schwarzschild radius

$$r_s = \frac{2GM}{c^2}, \tag{6.80}$$

and in terms of r_s the Schwarzschild metric signature becomes

$$ds^2 = -c^2 \, dt^2 \left(1 - \frac{r_s}{r}\right) + \frac{dr^2}{1 - r_s/r} + r^2 \, d\theta^2 + r^2 \sin^2 \theta \, d\phi^2. \tag{6.81}$$

The derivation of the Schwarzschild metric as a solution to the vacuum Einstein equations is a great example of the physicists intuition-first approach to solving complex equations. The solution reproduces the dynamics of Newton's law of gravitation for a spherically symmetric source outside that source. A more sophisticated treatment is necessary to understand gravity within the matter distribution of an extended spherically symmetric massive object, but it turns out that the Schwarzschild geometry is also applicable to such situations, with the mass M of the whole source replaced by the mass component $M(r)$ contained within a radius r. This is again in line with the characteristics of Newton's law of gravitation.

However, Schwarzschild brings us so much more than a geometric theory for the origins of Newton's law of gravity! As we will see in the next section, it brings us a model for one of the most iconic strange objects of twentieth century physics, the black hole.

6.9 The Event Horizon

Inspection of Equation (6.81) immediately reveals an apparent difficulty that is not a feature of Newtonian gravity. There is an apparent singularity in the metric signature at the Schwarzschild radius r_s. At this radius, ds^2 appears to diverge because of the factor of $1 - r_s/r$ in the denominator of the dr^2 term. It is not immediately clear what this means. You can see, however, that there is the possibility of some very interesting physics. There is no reason in the theory why a mass M cannot be concentrated in a spherically symmetric region of radius less than $2GM/c^2$. If it is, then an observer could in principle be present close to the Schwarzschild radius. What would happen?

For a body of the mass of the sun, the Schwarzschild radius is 3 km, far smaller than the solar radius of 7×10^5 km. It will therefore only be for objects far denser than our sun that we might expect to be able to study the strange physics that may occur close to $r = r_s$.

If the observer is in free fall, then the principle of equivalence has the answer. Nothing 'happens'. Such an observer cannot locally measure the gravitational field, and so there can be no special features of space-time at $r = r_s$. However, for an observer fixed against free fall close to $r = r_s$, things should be very different. This reminds us that we have to be very careful with the use of words from ordinary life, such as 'happen', when talking about black holes and other objects with strong gravity. The above argument strongly suggests that there is a very big difference between the events recorded by different observers in places where for observers fixed against gravity, a very strong gravitational field is predicted. We had better instead ask, more carefully, what would be measured by a particular observer in the vicinity of $r = r_s$ for the Schwarzschild solution.

One approach is to work out the trajectories of classes of freely falling bodies in our spherical coordinates (ct, r, θ, ϕ). There are several problems at the end of this chapter that carry out this programme in detail. The results have very important observational implications – the bending of light in a gravitational field and the precession of planetary orbits that before general relativity were thought to be elliptical amongst them. It is only because of lack of space that these problems are not part of the main text. Taken in small steps, however, these problems can be worked out by the student and will be very instructive in improving their understanding and providing valuable practice in the mathematical techniques.

The fact that a freely falling observer measures nothing special at $r = r_s$ means that though the mathematics superficially suggests a divergence in ds^2 in spherical polar coordinates, this cannot in reality be the case, because no such divergence is seen in the coordinate system of at least one observer, the freely falling one. Because ds^2 is an invariant, it must remain finite, even for the observer who is not in free fall. How is this possible, given Equation (6.81)? For a start, if we consider an observer at fixed r close to r_s, then for this observer, $dr = 0$, so the combination $dr^2/(1 - r_s/r)$ can in principle remain finite. All the same, things are extreme in polar coordinates. The presence of this singularity means that any attempt to integrate some physical trajectory along the path of some observer crossing $r = r_s$ will run into trouble because of the divergence. We might, however, hope to find a different set of coordinates in which behaviour of ds^2 at $r = r_s$ is easier to handle.

6.10 Kruskal–Szekeres Coordinates

Let us make a transformation from the (ct, r, θ, ϕ) spherical polar coordinate system by defining the new coordinates U and V, where

$$\left. \begin{aligned} U &= (\tfrac{r}{r_s} - 1)^{1/2} e^{r/(2r_s)} \cosh(\tfrac{ct}{2r_s}) \\ V &= (\tfrac{r}{r_s} - 1)^{1/2} e^{r/(2r_s)} \sinh(\tfrac{ct}{2r_s}) \end{aligned} \right\} r > r_s$$

$$\left. \begin{aligned} U &= (1 - \tfrac{r}{r_s})^{1/2} e^{r/(2r_s)} \sinh(\tfrac{ct}{2r_s}) \\ V &= (1 - \tfrac{r}{r_s})^{1/2} e^{r/(2r_s)} \cosh(\tfrac{ct}{2r_s}) \end{aligned} \right\} r < r_s. \tag{6.82}$$

The θ and ϕ coordinates are unchanged under the transformation, but we will in any case concentrate on radial trajectories, so that we will not need these coordinates.

Let us compute $U^2 - V^2$ in terms of r and t for $r > r_s$. We get

$$U^2 - V^2 = \left(\frac{r}{r_s} - 1\right) e^{r/r_s} \left(\cosh^2\left(\frac{ct}{2r_s}\right) - \sinh^2\left(\frac{ct}{2r_s}\right)\right)$$

$$= \left(\frac{r}{r_s} - 1\right) e^{r/r_s}. \tag{6.83}$$

Next, we compute $V^2 - U^2$ in terms of r and t for $r < r_s$. We get

$$V^2 - U^2 = \left(1 - \frac{r}{r_s}\right) e^{r/r_s} \left(\cosh^2\left(\frac{ct}{2r_s}\right) - \sinh^2\left(\frac{ct}{2r_s}\right)\right)$$

$$= \left(1 - \frac{r}{r_s}\right) e^{r/r_s}. \tag{6.84}$$

Now, if we fix r, then the right-hand side of both Equations (6.83) and (6.84), then the equations become very similar. Equation (6.83) becomes $U^2 - V^2 = P^2$, where P^2 is a positive constant. If you did Problem 2.1, then you will recognise this as the equation of a hyperbola with the two branches symmetric on either side of the V axis, intersecting with the U axis at $\pm P$. Equation (6.84) becomes $V^2 - U^2 = Q^2$, which similarly is a hyperbola but this time with the branches symmetric about the U axis, passing through the V axis at $\pm Q$.

These hyperbolae in this coordinate system correspond to surfaces of constant r. What about the point $r = 0$? This is at $r < r_s$, so we substitute $r = 0$ into Equation (6.84) and get $V^2 - U^2 = 1$. So in these coordinates, a single point in spherical coordinates at $r = 0$ is represented by every point on both branches of a hyperbola. Now consider the Schwarzschild radius $r = r_s$. Both Equations (6.83) and (6.84) yield $U = \pm V$, so that the Schwarzschild radius is represented by lines through the origin at $\pm 45°$ to the horizontal. Radii $r > r_s$ are represented by hyperbolae with the V axis as a symmetry axis passing through points $U =$

$\pm\sqrt{r/r_s - 1}\exp(r/r_s)$, which extends to ever larger U as r increases. All the radii between r_s and the origin are represented by hyperbolae with the U axis as the symmetry axis, that pass through the V axis at $V = \pm\sqrt{1 - r/r_s}\exp r/r_s$, which is between $V = 0$ for $r = r_s$ and $V = \pm 1$ for $r = 0$. Note that the space above the positive V branch of $V^2 - U^2 = 1$ and below the negative V branch of the same hyperbola does not have any real solution for r. So the whole space, all values of r, are represented by part of the UV plane, just as was the case in Problem 2.1.

Having found the lines of constant r, we figure out lines of constant t. Again following Problem 2.1, we take the ratio of the two equations. For $r > r_s$ and $r < r_s$, respectively, we get

$$\tanh\left(\frac{ct}{2r_s}\right) = \frac{V}{U} \qquad\qquad r > r_s$$

$$\tanh\left(\frac{ct}{2r_s}\right) = \frac{U}{V} \qquad\qquad r < r_s. \qquad (6.85)$$

In both cases, lines of constant t are represented by straight lines through the origin. Interestingly, at $r < r_s$, the gradient of the lines become greater for more positive t, moving from -1 for $t \to -\infty$ to $+1$ for $t \to +\infty$. For $r < r_s$, the gradients of the lines get less steep for more positive t. A portion of the UV plane with lines of constant r and t overlaid is shown in Figure 6.1. Perhaps, the oddest thing about this figure is that the point $r = 0$ becomes both branches of the hyperbola $V^2 - U^2 = 1$!

6.11 Metric Coefficients in Kruskal Coordinates

Let us figure out the paths taken by light rays, the null geodesics, in Kruskal–Szekeres coordinates. To do this, first figure out the metric ds^2. Let us do the case where $r < r_s$. The other case is similar and is left as an exercise for practice. From Equation (6.82) we have

$$U = \left(1 - \frac{r}{r_s}\right)^{1/2} e^{r/(2r_s)} \sinh\left(\frac{ct}{2r_s}\right)$$

$$V = \left(1 - \frac{r}{r_s}\right)^{1/2} e^{r/(2r_s)} \cosh\left(\frac{ct}{2r_s}\right). \qquad (6.86)$$

Some definitions will make it quicker to do the algebra:

$$T(t) = \frac{ct}{2r_s}$$

$$C(r) = \left(1 - \frac{r}{r_s}\right)^{1/2} e^{r/(2r_s)}. \qquad (6.87)$$

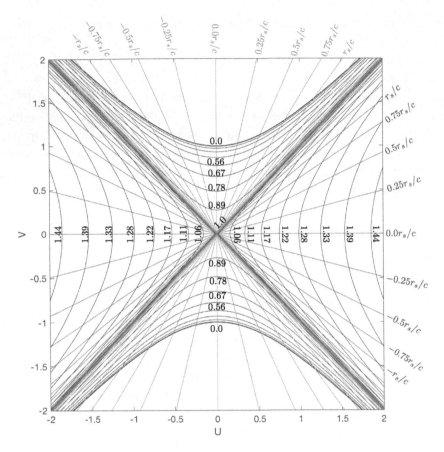

Figure 6.1 The UV plane with lines of constant r and t overlaid. The Schwarzschild radius corresponds to the lines $U = \pm V$. The hyperbolae correspond to
different radii as labelled, with distances from the origin labelled in units of r_s.
Lines of constant t are straight lines through the origin having gradient less than 1
at $r > r_s$ and greater than 1 at $r < r_s$. The time intervals between successive lines
are fixed, but the lines get more dense, and the density of the lines approaches
infinity as the Schwarzschild radius is approached.

Then write out sums and differences:

$$V - U = C(r)(\cosh T - \sinh T)$$
$$V + U = C(r)(\cosh T + \sinh T)$$
$$dV - dU = dC(r)(\cosh T - \sinh T) + C(r)(\sinh T - \cosh T)\,dT$$
$$dV + dU = dC(r)(\cosh T + \sinh T) + C(r)(\sinh T + \cosh T)\,dT$$
$$dV^2 - dU^2 = \left(dC(r)\right)^2 - C(r)^2\,dT^2$$
$$= \left(\frac{dC}{dr}\right)^2 dr^2 - C(r)^2\,dT^2. \tag{6.88}$$

This result also follows for $r > 2r_s$. Therefore the expression for $dV^2 - dU^2$ is independent of r.

The derivative dC/dr is

$$\frac{dC}{dr} = \frac{1}{2}\left(1 - \frac{r}{r_s}\right)^{-1/2}\frac{-1}{r_s}e^{r/(2r_s)} + \left(1 - \frac{r}{r_s}\right)^{1/2}\frac{1}{2r_s}e^{r/(2r_s)}$$

$$= e^{r/(2r_s)}\frac{-1}{2r_s}\left(\left(1 - \frac{r}{r_s}\right)^{1/2} - \frac{1}{(1 - r/r_s)^{1/2}}\right)$$

$$= \frac{r/(2r_s^2)e^{r/(2r_s)}}{(1 - r/r_s)^{1/2}}. \tag{6.89}$$

The differential dT is related to dt by

$$dT = \frac{c\,dt}{2r_s}. \tag{6.90}$$

Substituting $C(r)$, dC/dr, and dT into Equation (6.88), we obtain

$$dV^2 - dU^2 = \frac{r^2}{4r_s^4}e^{r/r_s}\frac{dr^2}{(1 - r/r_s)} - \left(1 - \frac{r_s}{r}\right)e^{r/r_s}\frac{c^2\,dt^2}{4r_s^2}$$

$$= \frac{r^2}{4r_s^4}e^{r/r_s}\frac{dr^2}{-(r/r_s)(1 - r_s/r)} + \frac{r}{r_s}\left(1 - \frac{r_s}{r}\right)e^{r/r_s}\frac{c^2\,dt^2}{4r_s^2}$$

$$= \frac{r}{4r_s^3}e^{r/r_s}\left(\frac{dr^2}{(1 - r_s/r)} - c^2\,dt^2\left(1 - \frac{r_s}{r}\right)\right). \tag{6.91}$$

Therefore

$$(-dV^2 + dU^2)\frac{4r_s^3}{r}e^{-r/(2r_s)} = -c^2\,dt^2\left(1 - \frac{r_s}{r}\right) + \frac{dr^2}{(1 - r_s/r)}. \tag{6.92}$$

We can therefore write the metric signature ds^2 as

$$ds^2 = (-dV^2 + dU^2)e^{-r/r_s}\frac{4r_s^3}{r} + r^2\,d\theta^2 + r^2\sin^2\theta\,d\phi^2. \tag{6.93}$$

Notice that ds^2 does not diverge at $r = r_s$ in these coordinates, confirming that the apparent singularity at $r = r_s$ in spherical polar coordinates is only due to the coordinate system stationary with respect to observers held stationary against free fall in the gravitational field of the black hole.

We can make a second very important observation from this expression. Consider photons on radial trajectories, for which $ds = 0$, and the paths are null geodesics. In this case, we must have

$$-dV^2 + dU^2 = 0$$

$$dV = \pm dU. \tag{6.94}$$

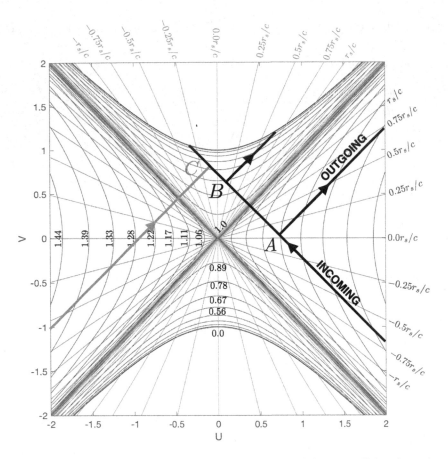

Figure 6.2 Space-time in Kruskal coordinates for particles on radial trajectories near to a black hole. Overlaid on the diagram in bold with arrows are world lines of some photons. Photons inside the Schwarzschild radius have their future geodesics inevitably terminating at $r = 0$, on the singularity. Photons outside the Schwarzschild radius have both ingoing plunge trajectories, where their world lines terminate at $r = 0$, and outgoing exit trajectories, where they eventually escape the singularity. Two photons originating on the left- and right-hand regions outside the Schwarzschild radius in the past can intersect at $r < r_s$, here at the point labelled C. Also shown are lines of constant time, which are straight lines through the origin at various gradients, and hyperboli at a range of constant radii.

This means that the world lines of photons are straight lines at 45° to the horizontal axis (of gradient ± 1) in the UV plane, just as they are on an ordinary space-time diagram in special relavitiy.

Figure 6.2 shows the important features of space-time for photons on radial trajectories in Kruskal coordinates near a black hole.

Also shown are lines of constant t, overlaid on the UV plane, and hyperbolae of constant r. Notice that at large U and small V, the hyperbolae intersect with the lines in a pattern that looks more and more like a rectangular grid as U increases. This reflects that space-time tends to flatness far from the black hole.

Some observations about the incoming photon. Before it reaches r_s, it must cross an infinite number of lines of constant t, since they get infinitely close together to the straight lines of slope 1 that represent the Schwarzschild radius. This means that to an observer making measurements in the spherical polar coordinate system, that is, having a network of clocks set up at different fixed radii along the pathway of the incoming photon, it would appear that the rate at which the photon travelled to smaller radii approaches zero as the photon approaches the Schwarzschild radius. This causes another feature that the crests and troughs of the photon have to get closer together, to bunch up, as the photon approaches r_s – it becomes shifted towards a shorter wavelength. Though it might seem strange that the photon has a dr/dt that is dependent on r, bear in mind that in the strong gravitational field of the black hole, the clocks and measurement apparatus in the spherical polar coordinate system are in a highly curved space-time, so that strange effects due to the coordinate system are expected.

In the freely falling frame of the photon, however, there is no such delay. The Schwarzschild radius is crossed without incident in a finite time, because to this photon, by the principle of equivalence, there is no measurable gravitational field and no space-time curvature. In fact, remembering that the photon has an infinite γ factor, any clock moving in the photons rest frame would stop because of time dilation, so that for a photon, all times are in fact equivalent anyway. Before the photon crosses $r = r_s$, were it to turn around and try and fly in the direction of increasing r, as is illustrated by the point B in Figure 6.2, the resulting photon world line does move to greater r, and the photon does escape the black hole, although if it starts to move outward from very close to r_s, then it may take a very long time to move away.

Once the photon has crossed to $r < r_s$, even if it tries to turn around and go in the other direction, as is illustrated at point B, both photon trajectories from this point are in fact still inwards and lead to the singularity at $r = 0$. So, no light can escape a black hole.

6.12 Wormholes

Notice from Figure 6.2 that only half the UV plane is needed to cover all space-time points. However, the metric ds^2 in Kruskal coordinates, including the whole plane, is an exact solution of Einstein's equations. The half–space represents the physics of a black hole. What does the whole space represent?

Notice that there are two regions where the lines of constant t and r form a grid that tends to a Cartesian grid, meaning that the space-time in these two regions is asymptotically flat. One of these regions is space-time far from the black hole. A hypothesis is that the full solution could be used to represent two flat space-times connected by a throat of admittedly highly distorted space-time close to $U = V = 0$. The throat representing the connecting region is called a wormhole. Also featuring on this space-time diagram is a second singularity at negative V. Nobody knows if there are physical manifestations of the entrances to wormholes in our Universe. We do think that there are physical manifestations of black holes, though, so we can speculate that wormholes also might exist.

Going back to our sketch of Kruskal–Szekeres space-time, is it possible to travel from our asymptotically flat region of the Universe to the other one through the wormhole? The answer is no, because the null geodesics define the causal past of any space-time point, so a world line cannot have a slope of less than 45° on this diagram. A photon can leave the singularity at negative V infinitely in the past at infinitely large positive U, and arrive infinitely in the future at infinitely large negative U, but there is no causally connected, or time-like, world line connecting the two asymptotically flat regions.

On the other hand, it is possible for world lines originating in the two different asymptotically flat regions in the past to intersect inside the Schwarzschild radius. We might speculate that this might mean that an observer passing inside the Schwarzschild radius close to a wormhole might, before perishing, be able to detect light from this second asymptotically flat region of space-time. This is illustrated by the world line having slope $+1$ that starts at negative U and intersects a second diagonal world line, which started at positive U at the point labelled C. This world line represents a photon from the left-hand asymptotically flat region intersecting our incoming photon from the asymptotically flat region on the right. So, if you ever have the misfortune to pass the event horizon of a black hole, then take a look over your shoulder, and maybe you will see light from another Universe before you perish!

6.13 Problems

6.1 The constant R_0 appeared when we integrated the equation $G_{00} = 0$ in Section 6.7, was evaluated by comparing with the Newtonian limit in Section 6.8, and was found to be negative, $R_0 = -r_s$. Consider the consequences for Equation (6.57). For some values of r, and you should specify the range, which particular desired feature laid out in Section 6.2 is not achieved, and why? If you know some complex analysis, then you might also work out a formula for the originally introduced constant Λ in this regime.

6.2 A metric for a charged non-spinning black hole can be written

$$ds^2 = -\left(1 - \frac{2GM}{rc^2} + \frac{Q^2 G}{4\pi\varepsilon_0 c^4 r^2}\right)c^2\,dt^2$$

$$+ \frac{dr^2}{(1 - 2GM/(rc^2) + Q^2 G/(4\pi\varepsilon_0 c^4 r^2))}$$

$$+ r^2\,d\theta^2 + r^2\sin^2\theta\,d\phi^2,$$

where M and Q are the mass and charge of the black hole, ε_0 is the permittivity of free space, and G is Newton's gravitational constant.

(a) Show that when $Q = 0$, this metric becomes the Schwarzschild metric.

(b) Show that ds^2 is infinite at $r = 0$ and at two other values of the radius, r_- and r_+, determined by the equations

$$r_- = \frac{A}{2} - \sqrt{\frac{A^2}{4} - B}$$

$$r_+ = \frac{A}{2} + \sqrt{\frac{A^2}{4} - B},$$

where A and B are functions of some or all of Q, G, M, and ε_0.

(c) Find the critical value of the charge-to-mass ratio Q/M of the black hole in terms of ε_0 and G, above which the real singularity at $r = 0$ is not excluded from the space-time at $r > r_+$. Demonstrate that your answer has the correct dimensions for the charge-to-mass ratio.

6.3 The Kerr metric for a spinning black hole is

$$ds^2 = -\frac{(\Delta - (a/c)^2\sin^2\theta)}{\rho^2}c^2\,dt^2 - 2\left(\frac{a}{c}\right)\frac{2GMr\sin^2\theta}{c^2\rho^2}c\,dt\,d\phi$$

$$+ \frac{(r^2 + (a/c)^2)^2 - (a/c)^2\Delta\sin^2\theta}{\rho^2}\sin^2\theta\,d\phi^2 + \frac{\rho^2}{\Delta}dr^2 + \rho^2\,d\theta^2,$$

where (r, θ, ϕ) are spherical polar coordinates, M is the mass of the black hole, a is the ratio of the black hole angular momentum to its mass, and the parameters Δ and ρ are defined by

$$\Delta = r^2 - \frac{2GMr}{c^2} + \left(\frac{a}{c}\right)^2$$

$$\rho^2 = r^2 + \left(\frac{a}{c}\right)^2\cos^2\theta.$$

(a) Show that this metric tends to the Schwarzschild metric as the black hole angular momentum tends to zero.

(b) Find an expression for a conserved quantity p_ϕ in terms of those quantities defined in the metric and show that far from the black hole, this conserved angular momentum tends to $mr^2 \sin^2\theta \, d\phi/d\tau$, where τ is the proper time.

(c) Find an expression for a conserved quantity p_0 in terms of the quantities defined in the Lagrangian.

(d) Express g^{tt}, $g^{t\phi}$, and $g^{\phi\phi}$ in terms of the metric coefficients.

(e) Derive a formula for p^ϕ in terms of p_ϕ, p_t, $g_{\phi\phi}$, g_{tt}, and $g_{t\phi}$.

6.4 The Schwarzschild metric for particles on radial trajectories is

$$ds^2 = -c^2 \, dt^2 \left(1 - \frac{r_s}{r}\right) + \frac{dr^2}{(1 - r_s/r)},$$

where ds^2 is the invariant interval, t is time, r is the radial displacement of the particle from the origin, $(c\,dt, dr)$ are small displacements in the coordinates r and t, and r_s is the Schwarzschild radius. The transformation to a modified set of coordinates (v, r) can be effected by making the following change of variables to replace ct by v:

$$ct = v - r - r_s \ln\left|\frac{r}{r_s} - 1\right|.$$

(a) Show that in these coordinates, ds^2 can be written

$$ds^2 = -\left(1 - \frac{r_s}{r}\right) dv^2 + 2\,dv\,dr.$$

(b) Write down a Lagrangian in terms of some or all of v, r, $dv/d\tau$, and $dr/d\tau$.

(c) Identify a cyclic coordinate and show that for all particle trajectories, we have

$$\frac{dr}{d\tau} - \left(1 - \frac{r_s}{r}\right)\frac{dv}{d\tau} = A,$$

where A is a constant.

(d) For photons, show that a class of trajectories exists where v is a constant. For these solutions, show that r must get smaller as t increases for $r > r_s$.

6.5 In this problem, we will figure out the scattering of a light beam in the gravitational field of a spherically symmetric body. This applies to any spherically symmetric body, not just black holes. The physical effect is called gravitational lensing. This is the main observational tool for probing dark matter at the scale of galaxy clusters and was used to effectively rule out massive compact halo objects (MACHOs) as a dominant component of galactic dark matter. We begin with Equation (6.81).

(a) We know that when a beam scatters off a spherically symmetric body, the entire trajectory of the beam is in a single plane that also contains the centre of mass of the body. Show that for a massless photon,

$$-c^2\,dt^2\left(1 - \frac{r_s}{r}\right) + \frac{dr^2}{1 - r_s/r} + r^2\,d\theta^2 = 0.$$

(b) Re-read Section 4.5 and then write down a Lagrangian for our photon propagating in the same plane as referred to in part (a).

(c) Identify two cyclic coordinates and the associated conserved quantities. Show that

$$\frac{d\theta}{dt} = \frac{(1 - r_s/r)Q}{r^2},$$

where Q is a constant.

(d) Hence show that

$$\frac{d\theta}{dr} = \frac{\pm 1}{r^2\sqrt{1/b^2 - (1/r^2)(1 - r_s/r)}},$$

where b is a constant.

(e) Consider the limit where the black hole mass tends to zero, so that $r_s = 2GM/c^2 = 0$. Show that in this limit, we get

$$\theta = \theta_0 \pm \arcsin\left(\frac{b}{r}\right).$$

(f) Show that this represents a particle passing the origin moving in a straight line and interpret the parameter b and the angle θ in terms of this solution. Illustrate your interpretations with a diagram of the trajectory as the light beam passes the origin. What are the two values of θ as $r \to \pm\infty$?

(g) Find a substitution such that in some new variable u, we have

$$d\theta = \frac{\pm du}{\sqrt{1/b^2 - u^2(1 - r_s u)}}.$$

(h) Now make a further substitution $w = u(1 - r_s u/2)$. Show making liberal use of the binomial theorem that, to first order in the small parameter r_s (small compared to r),

$$d\theta = \frac{\pm dw\,(1 + r_s w)}{\sqrt{1/b^2 - w^2}}.$$

(i) Integrate this equation to get θ as a function of w, r_s, and b.

(j) Show that, to first order in r_s, when you substitute back in for w to obtain $\theta(r)$, you get, to first order in r_s,

$$\theta(r) = \pm \arcsin\left(\frac{b}{r}\right) \mp \frac{r_s}{b}.$$

(k) Finally, show that as the beam travels past the origin where the massive body is located, it is scattered through an overall angle of

$$\Delta\theta = \frac{4GM}{bc^2}.$$

This is the relativistic formula for the scattering angle of light passing a massive body responsible for the phenomenon of gravitational lensing.

6.6 In this problem, we will use an effective potential approach to show that there are various classes of trajectories for bodies in free fall in the Schwarzschild geometry. Some of these are orbits approximating the elliptical orbits of Kepler and Newton, but others are entirely new and correspond to plunge trajectories into the black hole. The problem is in many parts and leads on to Problem 6.7 on the precession of the perihelion of Mercury.

(a) Show that the Lagrangian for a particle orbiting in a plane at $\theta = 90°$ about the Sun, considered as spherically symmetric, can be written

$$L = \sqrt{c^2\left(\frac{dt}{d\tau}\right)^2\left(1 - \frac{r_s}{r}\right) - \frac{(dr/d\tau)^2}{1 - r_s/r} - r^2\left(\frac{d\phi}{d\tau}\right)^2}, \qquad (6.95)$$

where (r, ϕ) and r_s are the coordinates of Mercury and the Schwarzschild radius of the sun, $r_s = 2GM_\odot/c^2$. If you calculate r_s, then it is only a few kilometres, far less than the solar radius.

(b) Find two conserved quantities v_α, where α is the coordinate corresponding to the conserved quantity. Multiply these conserved quantities by the mass m of Mercury, obtaining $p_\alpha = mv_\alpha$ for two different α. Raise the indices to obtain p^β corresponding to each of the conserved components.

(c) Using the conserved quantity $p_\alpha p^\alpha = -m^2c^2$, where the sum over α includes both the conserved and non-conserved components, show that

$$m^2\left(\frac{dr}{d\tau}\right)^2 = \frac{\tilde{E}^2}{c^2} - \left(\frac{\tilde{L}^2}{r^2} + m^2c^2\right)\left(1 - \frac{r_s}{r}\right). \qquad (6.96)$$

Identify $\tilde{L}$ and $\tilde{E}$ with your conserved quantities.

(d) Obtain an expression for $m^2(d\phi/d\tau)^2$ in terms of some or all of r, ϕ, $\tilde{L}$, and $\tilde{E}$.

(e) From classical mechanics we know that where the total energy of a particle E_{tot} is conserved, and the particle has potential energy V, then the difference between E_{tot} and V is the kinetic energy of the particle. The velocity of the particle is then $v = \sqrt{(2/m)(E_{tot} - V)}$. Often, V is a function of the particle position, so you can use this equation to work out how the velocity varies with position as the particle moves. Starting from Equation (6.96) and interpreting this equation as an expression for the radial velocity $dr/d\tau$ of a particle in terms of an effective total energy E_{tot} and an effective potential energy $V_{eff}(r)$, a function of the radial coordinate r, show that

$$E_{tot} = \frac{\tilde{E}^2}{2mc^2} \tag{6.97}$$

$$V_{eff}(r) = \frac{1}{2m}\left(\frac{\tilde{L}^2}{r^2} + m^2c^2\right)\left(1 - \frac{r_s}{r}\right). \tag{6.98}$$

(f) Multiply out Equation (6.98) and insert the formula for the Schwarzschild radius $r_s = 2GM/c^2$. Neglect the constant term, which is an overall additive offset connected to the rest energy of the orbiting satellite. From the remaining three terms identify the term yielding the Newtonian gravitational potential energy that dominates at large r, the term that imposes conservation of angular momentum, which in pre-relativity gravitation gives rise to stable elliptical orbits, and finally the term that dominates at small r, which allows bodies of sufficiently high total energy and sufficiently low angular momentum to approach $r = 0$. Make a sketch of $V_{eff}(r)$ and indicate on the sketch the radius that corresponds to a stable circular orbit. Also, mark on your sketch the Newtonian limit at large r and the value of the total effective energy that allows objects to pass the event horizon and fall into the black hole.

6.7 In this problem, we will work out a second famous result, this time the precession of the perihelion of Mercury. Historically, this was one of the early tests of the validity of general relativity. Kepler's laws, which are compatible with Newton's theory of gravitation, state that planetary orbits are elliptical. This means that they are closed; once an orbital period the planet's centre of mass returns to the same point. The star about which the planet rotates is at one focus of the ellipse, so that once an orbit the planet reaches its perihelion, or point of closest approach to the star. The position of the perihelion in Kepler's model remains the same.

General relativity, however, predicts that orbit of a planet is not closed and that therefore the perihelion undergoes precession. If we were to take the points of closest approach of the planet to the Sun in two subsequent orbits and measure the angle between straight lines drawn from the planet at these points to the Sun, we could in principle measure the precession in units of radians per orbit. It turns out, however, that the number is very small. It is customary instead to quote the calculated result in units of arc seconds (one arc second is 1/3600 of one degree) per century – the cumulative effect of the slow change in the position of closest approach over many hundreds of orbits! It should be emphasised that the general relativistic precession of the perihelion is just one of many corrections to the Keplerian approximation of elliptical orbits.

It is a testament to the ingenuity of the scientists involved that careful observations of Mercury's orbit had been made since the seventeen hundreds and that the data was successfully corrected for all these other effects! The residual precession of Mercury's perihelion was found to be in agreement with the prediction of general relativity.

The problem begins where Problem 6.6 left off. You are best off doing that problem as a prerequisite to this one.

(a) Show that $V_{\text{eff}}(r)$ in Equation (6.98) has a maximum at $r_1 = 3r_s/2$ and a minimum, neglecting a factor of order r_s, at $r_2 = 2L^2/(m^2c^2r_s)$. The latter radius corresponds to the radius of a stable circular orbit of Mercury about the Sun in the Newtonian/Keplerian approximation. In principle, were Mercury to be at that distance from the Sun, its orbit would be circular. In practice, firstly, Mercury is not at $r = r_2$, so even in the Newtonian/Keplerian approximation, Mercury's orbit is elliptical. Secondly, this is not quite the radius for a circular orbit; there is a correction of order r_s, which we have neglected for now. Thirdly, as alluded to in the introduction to this extended problem, the orbit of Mercury is not even quite elliptical when general relativistic effects are accounted for.

(b) Using the results from Problem 6.6, parts (c) and (d), show that

$$\tilde{L}^2\left(\frac{1}{r^2}\left(\frac{dr}{d\phi}\right)\right)^2 = \frac{\tilde{E}^2}{c^2} - \left(m^2c^2 + \frac{\tilde{L}^2}{r^2}\right)\left(1 - \frac{r_s}{r}\right). \qquad (6.99)$$

(c) Making the substitution $u = r_s/r$ into Equation (6.99), show that

$$\frac{1}{r_s^2}\left(\frac{du}{d\phi}\right)^2 = \frac{\tilde{E}^2 - m^2c^4}{\tilde{L}^2c^2} + \frac{m^2c^2u}{\tilde{L}^2} - \frac{u^2}{r_s^2} + \frac{u^3}{r_s^2}. \qquad (6.100)$$

(d) Neglecting the u^3 term in Equation (6.100) is making the assumption that $r \gg r_s$. Under this assumption, show that if we make the further substitution

$$y = u - \frac{(mcr_s)^2}{2\tilde{L}^2},$$

where the offset means that $y = 0$ corresponds to the radius of the circular orbit, then we obtain

$$\left(\frac{dy}{d\phi}\right)^2 = r_s^2 \left(\frac{\tilde{E}^2 - (mc^2)^2}{\tilde{L}^2 c^2} + \frac{(mcr_s)^2}{4\tilde{L}^4}\right) - y^2$$
$$= y_0^2 - y^2,$$
(6.101)

where y_0^2 is the combination of constants in the upper expression.

(e) By making an appropriate substitution integrate Equation (6.101) and show that $y(\phi)$ is periodic in ϕ with period 2π radians. This means that in this approximation the orbit returns to the same radius with the same period with which it returns to the same azimuthal angle. Therefore the orbit is closed; there is no precession, and this corresponds to the elliptical case observed by Kepler and modelled by Newton.

(f) Next, we return to Equation (6.100), but this time, we retain the u^3 term. Make the same substitution as before for y, but this time discard any terms that arise in y^3. Show that the resultant differential equation now takes the form

$$\left(\frac{dy}{d\phi}\right)^2 = y_0^2 + P^2 y - Q^2 y^2,$$
(6.102)

where P^2 and Q^2 are given by

$$P^2 = \frac{3}{4}\left(\frac{mcr_s}{\tilde{L}}\right)^4$$

$$Q^2 = 1 - \frac{3}{2}\left(\frac{mcr_s}{\tilde{L}}\right)^2.$$
(6.103)

(g) By completing the square in y on the right-hand side of Equation (6.102) show that it can be re-written as

$$\frac{dy}{d\phi} = \sqrt{H^2 - Q^2\left(y - \frac{P^2}{2Q^2}\right)^2},$$
(6.104)

where the choice of the positive square root corresponds to a choice of the direction of orbit of Mercury around the Sun, and the new constant H^2 is a function of y_0^2, P^2, and Q^2 to be determined.

(h) Integrate Equation (6.104), showing that once again, ϕ is periodic in y, but that in this more precise approximation, where we have neglected terms in y^3 rather than terms in u^3, the period differs from 2π by a correction term arising from Q and equal to $3\pi r_s/r$ radians, where r is the nominal orbital radius.

(i) Mercury's orbit is at nominal radius 5.55×10^{10} m. The period of the orbit is 0.24 years. The mass of the Sun is $M_\odot = 2.0 \times 10^{30}$ kg. Calculate the precession angle, first in radians per orbit, and then in arc seconds per century, hopefully arriving at 0.43 for the latter. This calculation was painful. The reality of calculations in general relativity almost always is, but this is a calculation of consequence! It was one of the first pieces of evidence that Einstein was on to something with his revolutionary geometric theory.

7

Cosmology

7.1 The Cosmological Principle

In the previous chapter, we set out to find the most general metric signature of a highly symmetric configuration of matter – one which was spherically symmetric and static. There is an enormous class of problems that conform more or less to these assumptions, from ball bearings to black holes.

We now turn to another highly symmetric problem. In this case, we are interested in the metric of a uniform, or homogeneous and isotropic space. This is a space where things are the same at every point at any given time, and, by extension, the same as observed in all directions from any given point. In a sense. it is the opposite of the previous problem. It turns out that in this case, we can also find an exact solution to Einstein's equations. This problem has one 'killer' application, the large-scale structure of the Universe. The assumption that on a sufficiently large scale the Universe is homogeneous and isotropic is precisely the cosmological principle on which modern cosmology and astrophysics are founded.

We will this time have non-zero stress energy tensor components; a Universe that is the same everywhere and contains zero energy density would be a boring place indeed. As a bonus, comparison with the results of a Newtonian analysis of a closely related problem will supply a numeric value for the constant A in Einstein's equations $G^{\alpha\beta} = AT^{\alpha\beta}$. As before, we will be interested in the most general metric signature that has the above-mentioned properties. This is the Friedmann–Lemaître–Robertson–Walker (FRW) metric. Our approach is slightly different from that in Chapter 6. We just write down the usual form of the metric, then we use that form to work out an expression for the Ricci scalar R. We then note that the scalar curvature is the same everywhere and that the metric therefore yields space-time compatible with the fundamental cosmological assumptions of isotropy and homogeneity. We then go on to derive the equations that govern the time evolution of the Universe. These equations are solved for some special cases in the problems at the end of the chapter.

7.2 Spaces of Uniform Curvature

We start with the Friedmann–Lemaître–Robertson–Walker metric signature

$$ds^2 = -c^2\,dt^2 + a(t)^2\left(\frac{dr^2}{1-kr^2} + r^2\,d\theta^2 + r^2\sin^2\theta\,d\phi^2\right). \tag{7.1}$$

The constant k can be either 0, $+1$, or -1. The parameter $a(t)$ is a dimensionless scale factor that can vary with time. However, it is not free to vary in any possible way; it is subject to dynamics governed by the matter and energy content of the Universe. Where the constant $k = 0$, the factor of $1 - kr^2$ becomes 1, and you are left with

$$\begin{aligned} ds^2 &= -c^2\,dt^2 + a(t)^2\left(dr^2 + r^2\,d\theta^2 + r^2\,\sin^2\theta\,d\phi^2\right) \\ &= -c^2\,dt^2 + a(t)^2\left(dx^2 + dy^2 + dz^2\right). \end{aligned} \tag{7.2}$$

Without the complicating influence of the kr^2 term, we can more clearly see the meaning of the $k = 0$ case, because we can re-write the spatial components in Cartesian form. The $k = 0$ case is the ordinary Euclidean space that uniformly expands and contracts.

7.3 Cosmologial Redshift

At time t, a photon travelling in the x direction has $ds^2 = 0$ and $dy = dz = 0$, so that

$$c^2\,dt^2 = a(t)^2\,dx^2$$

$$dx = \pm\frac{c\,dt}{a(t)}. \tag{7.3}$$

Now consider a photon making an extended journey along the x axis of a Universe governed by such a metric. Imagine it starts at time $t = 0$ and travels a distance L. Then we have

$$\int_{x=0}^{L} dx = c\int_{t=0}^{T}\frac{dt}{a(t)}. \tag{7.4}$$

Now a photon is not a point object. It is a series of wave fronts. The above expression is imagined to represent the passage of a node of the wave from point $x = 0$ to point $x = L$. Now the next node of the wave train starts out at time $t = T_i$, where T_i is the period of the wave at the moment it sets out, and arrives at time

$t = T + T_f$, where again T is the length of the journey of the first photon, and T_f is the period of the photon at the point where its journey ends. For this photon, the integral is

$$\int_{x=0}^{L} dx = c \int_{t=T_i}^{T+T_f} \frac{dt}{a(t)}.$$ (7.5)

We can equate the two right-hand sides, since they both integrate to the same quantity:

$$\int_{t=0}^{T} \frac{dt}{a(t)} = \int_{t=T_i}^{T+T_f} \frac{dt}{a(t)}.$$ (7.6)

We divide the left-hand integral into two portions, the first very short one between $t = 0$ and $t = T_i$, and the second far longer one between $t = T_i$ and $t = T$. We divide the right-hand integral also between two portions, the first between $t = T_i$ and $t = T$, and the second between $t = T$ and $t = T + T_f$. We obtain

$$\int_{t=0}^{T_i} \frac{dt}{a(t)} + \int_{t=T_i}^{T} \frac{dt}{a(t)} = \int_{t=T_i}^{T} \frac{dt}{a(t)} + \int_{t=T}^{T+T_f} \frac{dt}{a(t)}.$$ (7.7)

We cancel the integral in common, so that

$$\int_{t=0}^{T_i} \frac{dt}{a(t)} = \int_{t=T}^{T+T_f} \frac{dt}{a(t)}.$$ (7.8)

These integrals are each over only a single period of the wave. Because over this tiny interval $a(t)$, the cosmic scale factor, is not going to change very much, we can factor out the $a(t)$ and call it $a(t_i)$, the scale factor at the time of emission. Similarly, we can factor out the $a(t)$ from the right-hand integral and call it $a(t_f)$, the scale factor at the time of absorption. The integral now just becomes the difference between the time limits, so we have

$$\frac{T_i}{a(t_i)} = \frac{T_f}{a(t_f)}.$$ (7.9)

The period of the wave is just λ/c, where λ is the wavelength of the wave, and hence

$$\frac{\lambda_f}{\lambda_i} = \frac{a(t_f)}{a(t_i)}.$$ (7.10)

This is the cosmological redshift in action. The scale factor $a(t)$ causes the entire Universe to 'breathe', to rarify and compress in uniform, always maintaining isotropy. As it does so, the wavelengths of all the photons in the Universe simultaneously scale up and down. Since the big bang, we know from observations that the

Universe has been expanding, and so along with it the wavelengths of all the photons in the Universe have been getting larger, so those photons have been increasing in wavelength and hence losing energy. This flagrantly violates our expectation that energy is overall conserved, but what right do we have to expect global conservation of energy over the entire Universe? After all, what coordinate system can be built that encompasses the whole Universe? There can only be local coordinate systems covering small patches. There is in fact no global conservation of energy implied by the FRW metric. Indeed, time is not a cyclic coordinate, so even in that we can see that energy conservation is going to be violated.

Notice that this same argument would also work in the case of $k = \pm 1$ because the spatial integral would disappear from the calculation in the same way if there was $1 - kr^2$ in the denominator and the photon was travelling in the direction of increasing r. So, redshift of photons also shows up in universes with $k = \pm 1$, whatever they turn out to represent.

7.4 What Is the r Coordinate?

Looking again at Equation (7.1), this does in fact seem a little peculiar. In particular, the denominator of the term in dr^2 is $1 - kr^2$. We have said that k can be either 0 or ± 1. The problem is that when it is ± 1, the term kr^2 is dimensionally incompatible with the 1 if r is interpreted as a radius in metres. In these cases, we must conclude that the symbol r does not represent a number with the dimensions of length. The resolution of this apparent paradox is that with $k \neq 0$, then the metric will turn out to represent non-Euclidean spaces.

In fact, with $k = +1$, the metric represents the three-dimensional analog of a spherical surface, a three-dimensional hyperspherical 'surface'. On a sphere, both of the two coordinates that you need to describe your position are angles. Similarly, on the surface of a hypersphere, all three of the coordinates used to describe your position are angles, including the r coordinate. If you are uncomfortable with this, then consider that a hypersphere has a radius too, and the distance that you travel in the direction of increasing r, which is in the hyperspherical surface, divided by the radius of the hypersphere (which is a line normal to the hypersurface), can be thought of as an angle in radians and therefore dimensionless. A similar argument can be made for the r coordinate when $k = -1$, and it turns out that the metric represents a three-dimensional space of constant negative curvature. Such a space has no two-dimensional analog (a saddle has negative curvature, but it is not the same curvature everywhere)

We will go on to prove all of this by working out the geometry of the FRW metric, again by calculating out all the tensor components of the curvature and

Einstein tensors, and then setting the Einstein tensor coefficients proportional to the components of the stress–energy tensor.

7.5 Lagrangian and Christoffel Symbols

To figure out the equations that govern $a(t)$ and hence determine the expansion history of the Universe, we need to find the components of the Einstein tensor associated with the FRW metric. We start with the resulting Lagrangian

$$L = \left[\left(\frac{d(ct)}{d\tau} \right)^2 - \frac{a^2}{1-kr^2} \left(\frac{dr}{d\tau} \right)^2 \right.$$
$$\left. - a^2r^2 \left(\frac{d\theta}{d\tau} \right)^2 - a^2r^2 \sin^2\theta \left(\frac{d\phi}{d\tau} \right)^2 \right]^{1/2}. \tag{7.11}$$

The partial derivative with respect to $d(ct)/d\tau$ is

$$\frac{\partial L}{\partial(d(ct)/d\tau)} = \frac{1}{2L} 2 \frac{d(ct)}{d\tau} = \frac{1}{L} \frac{d(ct)}{d\tau}. \tag{7.12}$$

Its derivative with respect to τ, recalling that $dL/d\tau = 0$, is

$$\frac{d}{d\tau} \frac{\partial L}{\partial(d(ct)/d\tau)} = \frac{1}{L} \frac{d^2(ct)}{d\tau^2}. \tag{7.13}$$

The partial derivative of L with respect to ct is

$$\frac{1}{c} \frac{\partial L}{\partial t} = \frac{1}{L} \left(\frac{-a\dot{a}}{c(1-kr^2)} \left(\frac{dr}{d\tau} \right)^2 - \frac{a\dot{a}r^2}{c} \left(\frac{d\theta}{d\tau} \right)^2 - \frac{a\dot{a}r^2 \sin^2\theta}{c} \left(\frac{d\phi}{d\tau} \right)^2 \right), \tag{7.14}$$

where $\dot{a} = da/dt$. The Euler–Lagrange equation for the ct degree of freedom is

$$\frac{d^2(ct)}{d\tau^2} + \frac{a\dot{a}}{c(1-kr^2)} \left(\frac{dr}{d\tau} \right)^2 + \frac{a\dot{a}r^2}{c} \left(\frac{d\theta}{d\tau} \right)^2 + \frac{a\dot{a}r^2 \sin^2\theta}{c} \left(\frac{d\phi}{d\tau} \right)^2 = 0. \tag{7.15}$$

We read off the first three non-zero Christoffel symbols:

$$\Gamma^0_{11} = \frac{a\dot{a}}{c(1-kr^2)}$$

$$\Gamma^0_{22} = \frac{a\dot{a}r^2}{c}$$

$$\Gamma^0_{33} = \frac{a\dot{a}r^2 \sin^2\theta}{c}. \tag{7.16}$$

The partial derivative of L with respect to $dr/d\tau$ is

$$\frac{\partial L}{\partial(dr/d\tau)} = -\frac{1}{L} \frac{a^2}{1-kr^2} \frac{dr}{d\tau}, \tag{7.17}$$

and its τ derivative is

$$\frac{d}{d\tau}\frac{\partial L}{\partial(dr/d\tau)} = -\frac{1}{L}\left(\frac{a^2}{1-kr^2}\frac{d^2r}{d\tau^2} + \frac{2a\dot{a}}{c}\frac{1}{1-kr^2}\frac{d(ct)}{d\tau}\frac{dr}{d\tau}\right.$$

$$\left. + \frac{2kra^2}{(1-kr^2)^2}\left(\frac{dr}{d\tau}\right)^2\right). \tag{7.18}$$

The partial derivative of L with respect to r is

$$\frac{\partial L}{\partial r} = \frac{1}{L}\left(\frac{-kra^2}{(1-kr^2)^2}\left(\frac{dr}{d\tau}\right)^2 - a^2r\left(\frac{d\theta}{d\tau}\right)^2 - a^2r\sin^2\theta\left(\frac{d\phi}{d\tau}\right)^2\right). \tag{7.19}$$

The Euler–Lagrange equation for the r coordinate is

$$\frac{d^2r}{d\tau^2} + \frac{2\dot{a}}{ca}\frac{d(ct)}{d\tau}\frac{dr}{d\tau} + \frac{kr}{1-kr^2}\left(\frac{dr}{d\tau}\right)^2$$

$$- r(1-kr^2)\left(\frac{d\theta}{d\tau}\right)^2 - r(1-kr^2)\sin^2\theta\left(\frac{d\phi}{d\tau}\right)^2 = 0. \tag{7.20}$$

We infer the next five non-zero Christoffel symbols:

$$\Gamma^1_{01} = \Gamma^1_{10} = \frac{\dot{a}}{ca}$$

$$\Gamma^1_{11} = \frac{kr}{1-kr^2}$$

$$\Gamma^1_{22} = -r(1-kr^2)$$

$$\Gamma^1_{33} = -r(1-kr^2)\sin^2\theta. \tag{7.21}$$

The partial derivative of L with respect to $d\theta/d\tau$ is

$$\frac{\partial L}{\partial(d\theta/d\tau)} = -\frac{1}{L}a^2r^2\frac{d\theta}{d\tau}, \tag{7.22}$$

and its derivative with respect to τ is

$$\frac{d}{d\tau}\frac{\partial L}{\partial(d\theta/d\tau)} = -\frac{1}{L}\left(a^2r^2\frac{d^2\theta}{d\tau^2} + \frac{2a\dot{a}r^2}{c}\frac{d(ct)}{d\tau}\frac{d\theta}{d\tau} + 2a^2r\frac{dr}{d\tau}\frac{d\theta}{d\tau}\right). \tag{7.23}$$

The partial derivative of L with respect to θ is

$$\frac{\partial L}{\partial\theta} = -\frac{1}{L}a^2r^2\sin\theta\cos\theta\left(\frac{d\phi}{d\tau}\right)^2. \tag{7.24}$$

The Euler–Lagrange equation for the θ component is

$$\frac{d^2\theta}{d\tau^2} + \frac{2\dot{a}}{ca}\frac{d(ct)}{d\tau}\frac{d\theta}{d\tau} + \frac{2}{r}\frac{dr}{d\tau}\frac{d\theta}{d\tau} - \sin\theta\cos\theta\left(\frac{d\phi}{d\tau}\right)^2 = 0, \tag{7.25}$$

from which we read off the next five non-zero Christoffel symbols:

$$\Gamma^2_{02} = \Gamma^2_{20} = \frac{\dot{a}}{ca}$$

$$\Gamma^2_{12} = \Gamma^2_{21} = \frac{1}{r}$$

$$\Gamma^2_{33} = -\sin\theta\cos\theta. \tag{7.26}$$

The partial derivative of L with respect to $d\phi/d\tau$ is

$$\frac{\partial L}{\partial(d\phi/d\tau)} = -\frac{1}{L}a^2 r^2 \sin^2\theta \frac{d\phi}{d\tau}, \tag{7.27}$$

and its derivative with respect to τ is

$$\frac{d}{d\tau}\frac{\partial L}{\partial(d\phi/d\tau)} = -\frac{1}{L}\left(a^2 r^2 \sin^2\theta \frac{d^2\phi}{d\tau^2} + 2a\frac{da}{d\tau}r^2 \sin^2\theta \frac{d\phi}{d\tau}\right.$$

$$\left. + 2ra^2 \sin^2\theta \frac{dr}{d\tau}\frac{d\phi}{d\tau} + a^2 r^2 2\sin\theta\cos\theta \frac{d\theta}{d\tau}\frac{d\phi}{d\tau}\right), \tag{7.28}$$

and since the partial derivative of L with respect to ϕ is zero, the Euler–Lagrange equation for the ϕ coordinate is

$$\frac{d^2\phi}{d\tau^2} + \frac{2\dot{a}}{ca}\frac{d(ct)}{d\tau}\frac{d\phi}{d\tau} + \frac{2}{r}\frac{dr}{d\tau}\frac{d\phi}{d\tau} + \frac{2}{\tan\theta}\frac{d\theta}{d\tau}\frac{d\phi}{d\tau} = 0. \tag{7.29}$$

We can now list all 19 non-zero Christoffel symbols in the FRW geometry:

$$\Gamma^0_{11} = \frac{a\dot{a}}{c(1-kr^2)}$$

$$\Gamma^0_{22} = \frac{a\dot{a}r^2}{c}$$

$$\Gamma^0_{33} = \frac{a\dot{a}r^2 \sin^2\theta}{c}$$

$$\Gamma^1_{01} = \Gamma^1_{10} = \frac{\dot{a}}{ca}$$

$$\Gamma^1_{11} = \frac{kr}{1-kr^2}$$

$$\Gamma^1_{22} = -r(1-kr^2)$$

$$\Gamma^1_{33} = -r(1-kr^2)\sin^2\theta$$

$$\Gamma^2_{02} = \Gamma^2_{20} = \frac{\dot{a}}{ca}$$

$$\Gamma^2_{12} = \Gamma^2_{21} = \frac{1}{r}$$

$$\Gamma^2_{33} = -\sin\theta\cos\theta$$

$$\Gamma^3_{03} = \Gamma^3_{30} = \frac{\dot{a}}{ca}$$

$$\Gamma^3_{13} = \Gamma^3_{31} = \frac{1}{r}$$

$$\Gamma^3_{23} = \Gamma^3_{32} = \frac{1}{\tan\theta}. \tag{7.30}$$

We now go on to calculate the non-zero coefficients of the Riemann tensors.

7.6 Riemann Curvature

The 20 independent potentially non-zero Riemann curvature coefficients are $R^0{}_{101}$, $R^0{}_{202}$, $R^0{}_{303}$, $R^1{}_{212}$, $R^1{}_{313}$, $R^2{}_{323}$, $R^0{}_{123}$, $R^0{}_{213}$, $R^0{}_{102}$, $R^0{}_{103}$, $R^0{}_{203}$, $R^1{}_{012}$, $R^1{}_{013}$, $R^1{}_{213}$, $R^2{}_{021}$, $R^2{}_{023}$, $R^2{}_{123}$, $R^3{}_{031}$, $R^3{}_{032}$, and $R^3{}_{132}$.

The procedure for evaluating these components is the same as outlined in Section 6.4. As for the Schwarzschild geometry, it turns out that only the components with two pairs of two identical indices are non-zero. The student is left to show that the components with the indices are all the same, those with two indices are the same, and the other two different from this pair and from each other are all zero. The six independent non-zero components are

$$
\begin{aligned}
R^0{}_{101} &= \frac{\partial\Gamma^0_{11}}{\partial x^0} - \frac{\partial\Gamma^0_{01}}{\partial x^1} + \Gamma^0_{00}\Gamma^0_{11} + \Gamma^0_{01}\Gamma^1_{11} + \Gamma^0_{02}\Gamma^2_{11} + \Gamma^0_{03}\Gamma^3_{11} \\
&\quad - \Gamma^0_{10}\Gamma^0_{01} - \Gamma^0_{11}\Gamma^1_{01} - \Gamma^0_{12}\Gamma^2_{01} - \Gamma^0_{13}\Gamma^3_{01} \\
&= \frac{1}{c}\frac{\partial}{\partial t}\frac{a\dot{a}}{c(1-kr^2)} + \frac{\dot{a}a}{c(1-kr^2)}\frac{\dot{a}}{ca} = \frac{a\ddot{a}}{c^2(1-kr^2)},
\end{aligned}
\tag{7.31}
$$

$$
\begin{aligned}
R^0{}_{202} &= \frac{\partial\Gamma^0_{22}}{\partial x^0} - \frac{\partial\Gamma^0_{02}}{\partial x^2} + \Gamma^0_{00}\Gamma^0_{22} + \Gamma^0_{01}\Gamma^1_{22} + \Gamma^0_{02}\Gamma^2_{22} + \Gamma^0_{03}\Gamma^3_{22} \\
&\quad - \Gamma^0_{20}\Gamma^0_{02} - \Gamma^0_{21}\Gamma^1_{02} - \Gamma^0_{22}\Gamma^2_{02} - \Gamma^0_{23}\Gamma^3_{02} \\
&= \frac{1}{c}\frac{\partial}{\partial t}\frac{a\dot{a}r^2}{c} - \frac{r^2\dot{a}^2}{c^2} = \frac{a\ddot{a}r^2}{c^2},
\end{aligned}
\tag{7.32}
$$

$$
\begin{aligned}
R^0{}_{303} &= \frac{\partial\Gamma^0_{33}}{\partial x^0} - \frac{\partial\Gamma^0_{03}}{\partial x^3} + \Gamma^0_{00}\Gamma^0_{33} + \Gamma^0_{01}\Gamma^1_{33} + \Gamma^0_{02}\Gamma^2_{33} + \Gamma^0_{03}\Gamma^3_{33} \\
&\quad - \Gamma^0_{30}\Gamma^0_{03} - \Gamma^0_{31}\Gamma^1_{03} - \Gamma^0_{32}\Gamma^2_{03} - \Gamma^0_{33}\Gamma^3_{03} \\
&= \frac{1}{c}\frac{\partial}{\partial t}\left(\frac{a\dot{a}r^2\sin^2\theta}{c}\right) - \frac{a\dot{a}r^2\sin^2\theta}{c}\frac{\dot{a}}{ca} = \frac{a\ddot{a}r^2\sin^2\theta}{c^2},
\end{aligned}
\tag{7.33}
$$

$$R^1{}_{212} = \frac{\partial \Gamma^1_{22}}{\partial x^1} - \frac{\partial \Gamma^1_{12}}{\partial x^2} + \Gamma^1_{10}\Gamma^0_{22} + \Gamma^1_{11}\Gamma^1_{22} + \Gamma^1_{12}\Gamma^2_{22} + \Gamma^1_{13}\Gamma^3_{22}$$

$$- \Gamma^1_{20}\Gamma^0_{12} - \Gamma^1_{21}\Gamma^1_{12} - \Gamma^1_{22}\Gamma^2_{12} - \Gamma^1_{23}\Gamma^3_{12}$$

$$= \frac{\partial}{\partial r}\left(-r\left(1 - kr^2\right)\right) + \frac{\dot{a}}{ca}\frac{a\dot{a}r^2}{c} + \frac{kr}{1 - kr^2}\left(-r\left(1 - kr^2\right)\right) + \frac{r(1 - kr^2)}{r}$$

$$= \left(k + \frac{\dot{a}^2}{c^2}\right)r^2, \tag{7.34}$$

$$R^1{}_{313} = \frac{\partial \Gamma^1_{33}}{\partial x^1} - \frac{\partial \Gamma^1_{13}}{\partial x^3} + \Gamma^1_{10}\Gamma^0_{33} + \Gamma^1_{11}\Gamma^1_{33} + \Gamma^1_{12}\Gamma^2_{33} + \Gamma^1_{13}\Gamma^3_{33}$$

$$- \Gamma^1_{30}\Gamma^0_{13} - \Gamma^1_{31}\Gamma^1_{13} - \Gamma^1_{32}\Gamma^2_{13} - \Gamma^1_{33}\Gamma^3_{13}$$

$$= \frac{\partial}{\partial r}\left(-r\left(1 - kr^2\right)\sin^2\theta\right) + \frac{\dot{a}^2 r^2 \sin^2\theta}{c^2}$$

$$+ \frac{kr}{1 - kr^2}\left(-r\left(1 - kr^2\right)\sin^2\theta\right) - \left(\frac{-r(1 - kr^2)\sin^2\theta}{r}\right)$$

$$= \left(k + \frac{\dot{a}^2}{c^2}\right)r^2 \sin^2\theta, \tag{7.35}$$

$$R^2{}_{323} = \frac{\partial \Gamma^2_{33}}{\partial x^2} - \frac{\partial \Gamma^2_{23}}{\partial x^3} + \Gamma^2_{20}\Gamma^0_{33} + \Gamma^2_{21}\Gamma^1_{33} + \Gamma^2_{22}\Gamma^2_{33} + \Gamma^2_{23}\Gamma^3_{33}$$

$$- \Gamma^2_{30}\Gamma^0_{23} - \Gamma^2_{31}\Gamma^1_{23} - \Gamma^2_{32}\Gamma^2_{23} - \Gamma^2_{33}\Gamma^3_{23}$$

$$= \frac{\partial}{\partial \theta}\left(-\sin\theta\cos\theta\right) + \frac{\dot{a}^2 r^2 \sin^2\theta}{c^2} - \left(1 - kr^2\right)\sin^2\theta + \cos^2\theta$$

$$= \left(k + \frac{\dot{a}^2}{c^2}\right)r^2 \sin^2\theta. \tag{7.36}$$

7.7 Ricci Tensor Coefficients

The Ricci tensor coefficients can be determined following the procedure outlined in Section 6.5. All off-diagonal coefficients are zero because they contain only contributions from Riemann coefficients having three distinct indices, all of which are zero. For the non-zero diagonal Ricci components, all contributions containing Riemann coefficients with all indices equal are zero and are therefore omitted from the calculations. The symmetry of the Riemann coefficients $R_{\beta\alpha\delta\gamma} = R_{\alpha\beta\gamma\delta}$ is used to express all the Riemann coefficients that appear in terms of the six we have already calculated in Section 7.6. So,

$$R_{00} = R^1{}_{010} + R^2{}_{020} + R^3{}_{030}$$

$$= g^{11}g_{00}R^0{}_{101} + g^{22}g_{00}R^0{}_{202} + g^{33}g_{00}R^0{}_{303} = \frac{-3\ddot{a}}{c^2 a}, \tag{7.37}$$

$$R_{11} = R^0{}_{101} + R^2{}_{121} + R^3{}_{131}$$

$$= R^0{}_{101} + g^{22}g_{11}R^1{}_{212} + g^{33}g_{11}R^1{}_{313}$$

$$= \frac{a\ddot{a}}{c^2(1 - kr^2)} + \frac{2(k + \dot{a}^2/c^2)}{1 - kr^2}, \tag{7.38}$$

$$R_{22} = R^0{}_{202} + R^1{}_{212} + R^3{}_{232}$$

$$= R^0{}_{202} + R^1{}_{212} + g^{33}g_{22}R^2{}_{323} = \frac{a\ddot{a}r^2}{c^2} + 2\left(k + \frac{\dot{a}^2}{c^2}\right)r^2, \tag{7.39}$$

$$R_{33} = R^0{}_{303} + R^1{}_{313} + R^2{}_{323} = \frac{a\ddot{a}r^2 \sin^2\theta}{c^2} + 2\left(k + \frac{\dot{a}^2}{c^2}\right)r^2 \sin^2\theta. \tag{7.40}$$

7.8 The Ricci Scalar

The Ricci scalar is calculated from the Ricci tensor coefficients using $R = g^{\alpha\beta}R_{\alpha\beta}$:

$$R = g^{00}R_{00} + g^{11}R_{11} + g^{22}R_{22} + g^{33}R_{33}$$

$$= \frac{3\ddot{a}}{c^2 a} + \frac{1 - kr^2}{a^2}\left(\frac{a\ddot{a}}{c^2(1 - kr^2)} + \frac{2(k + \dot{a}^2/c^2)}{1 - kr^2}\right)$$

$$+ \frac{1}{a^2 r^2}\left(\frac{a\ddot{a}r^2}{c^2} + 2r^2\left(k + \frac{\dot{a}^2}{c^2}\right)\right)$$

$$+ \frac{1}{a^2 r^2 \sin^2\theta}\left(\frac{a\ddot{a}r^2 \sin^2\theta}{c^2} + 2r^2 \sin^2\theta\left(k + \frac{\dot{a}^2}{c^2}\right)\right)$$

$$R = 6\left(\frac{\ddot{a}}{c^2 a} + \frac{\dot{a}^2}{c^2 a^2} + \frac{k}{a^2}\right). \tag{7.41}$$

From this result we can see the meaning of the parameter k. Consider a Universe momentarily at a point of inflection in its scale evolution, so that $\dot{a} = \ddot{a} = 0$. In this case the Universe is spatially static. If $k = 0$, then the Ricci scalar is zero, and there is no scalar curvature anywhere. If $k = \pm1$, then the scalar curvature is $\pm 6/a^2$, which is a static positive (negative) curvature in the case of the plus (minus) sign, the same everywhere in the Universe, and proportional to one over the square of the scale factor. This confirms that $k = \pm1$ correspond to closed (plus sign implies positive curvature) and open (minus sign implies negative curvature) Universes obeying the cosmological principal that the curvature is position independent. Spaces of constant positive and negative curvature are called de Sitter and anti-de Sitter spaces, respectively.

7.9 The Einstein Tensor

Because the metric signature and the Ricci tensor components are both diagonal, only the diagonal components of the Einstein tensor are non-zero. The zero-zero component is

$$G_{00} = R_{00} - \frac{1}{2}g_{00}R$$

$$= \frac{-3\ddot{a}}{c^2 a} + \frac{3\ddot{a}}{c^2 a} + \frac{3\dot{a}^2}{c^2 a^2} + \frac{3k}{a^2}$$

$$G_{00} = \frac{3}{c^2}\left(\frac{\dot{a}}{a}\right)^2 + \frac{3k}{a^2}$$

$$G^{00} = g^{00}g^{00}G_{00} = G_{00}, \tag{7.42}$$

$$G_{11} = R_{11} - \frac{1}{2}g_{11}R$$

$$= \frac{a\ddot{a}}{c^2(1-kr^2)} - \frac{1}{2}\frac{a^2}{1-kr^2}6\left(\frac{\ddot{a}}{c^2 a} + \frac{\dot{a}^2}{c^2 a^2} + \frac{k}{a^2}\right)$$

$$= \frac{-a^2}{c^2(1-kr^2)}\left(\frac{2\ddot{a}}{a} + \left(\frac{\dot{a}}{a}\right)^2 + \frac{kc^2}{a^2}\right)$$

$$G^{11} = g^{11}g^{11}G_{11} = -\frac{(1-kr^2)}{c^2 a^2}\left(\frac{2\ddot{a}}{a} + \left(\frac{\dot{a}}{a}\right)^2 + \frac{kc^2}{a^2}\right), \tag{7.43}$$

$$G_{22} = R_{22} - \frac{1}{2}g_{22}R$$

$$= \frac{-a^2 r^2}{c^2}\left(\frac{2\ddot{a}}{a} + \left(\frac{\dot{a}}{a}\right)^2 + \frac{kc^2}{a^2}\right)$$

$$G^{22} = g^{22}g^{22}G_{22} = \frac{-1}{c^2 a^2 r^2}\left(\frac{2\ddot{a}}{a} + \left(\frac{\dot{a}}{a}\right)^2 + \frac{kc^2}{a^2}\right), \tag{7.44}$$

$$G_{33} = R_{33} - \frac{1}{2}g_{33}R$$

$$= \frac{-a^2 r^2 \sin^2\theta}{c^2}\left(\frac{2\ddot{a}}{a} + \left(\frac{\dot{a}}{a}\right)^2 + \frac{kc^2}{a^2}\right)$$

$$G^{33} = g^{33}g^{33}G_{33} = \frac{-1}{c^2 a^2 r^2 \sin^2\theta}\left(\frac{2\ddot{a}}{a} + \left(\frac{\dot{a}}{a}\right)^2 + \frac{kc^2}{a^2}\right). \tag{7.45}$$

7.10 Einstein's Equations

We now begin to substitute the components of the Einstein tensor into Einstein's equations and see what we learn. The 00 component yields $G^{00} = AT^{00} = A\rho$, where ρ is the energy density. We obtain

$$\frac{3}{c^2}\left(\frac{\dot{a}}{a}\right)^2 + \frac{3k}{a^2} = A\rho. \tag{7.46}$$

The quantity ρ represents the energy density of all contributing substances combined – dust, relativistic particles, electromagnetic radiation, and anything else that might contribute.

To evaluate the unknown constant A, we again compare the relativistic result with a naive calculation applicable in Euclidean space to a uniformly dense distribution of dust. This treatment considers a sphere of matter having uniform mass density, of radius $a(t)R$, where $a(t)$ is the same dimensionless scaling factor, and R is a fixed radius scale. We consider the case where $k = 0$, so that this body is in a Euclidean space. A body of mass m on the surface of this sphere is falling inwards under the gravitational attraction of the rest of the matter in the sphere such that its kinetic energy plus the gravitational potential energy of the body with respect to the rest of the matter in the sphere is zero. All of its kinetic energy can be considered as having been gained as it fell from infinity to form a particle on the boundary of this spherically symmetric distribution of matter.

Of course, this model is not a realistic cosmology. In the FRW cosmology the particle is also under the influence of all the other matter in the Universe. However, there is no reason why the FRW metric cannot be locally valid anywhere inside a spherically symmetric uniform matter distribution with an edge, because Einstein's equations are local. Because both the FRW and the Newtonian analysis can be performed on the same finite matter distribution, their results can be compared.

In the Newtonian analysis, therefore, the gravitational potential energy of the test particle in the gravitational field of the sphere of matter is the negative of its kinetic energy. So, we can write

$$\frac{1}{2}mR^2\dot{a}^2 = \frac{G((4/3)\pi R^3 a^3(\rho/c^2))m}{aR}, \tag{7.47}$$

where the brackets on the right contain the total mass of the contents of the sphere, and the negative of the right-hand side is the gravitational potential energy of the test mass m in the gravitational field of the spherical mass distribution of energy density ρ and hence of mass density ρ/c^2. Thus

$$\left(\frac{\dot{a}}{a}\right)^2 = \frac{8\pi G\rho}{3c^2}. \tag{7.48}$$

Now let us compare this with the equivalent result derived from Einstein's equations. We set $k = 0$ because the analysis is only compatible in Euclidean space. Equation (7.46) yields

$$\left(\frac{\dot{a}}{a}\right)^2 = \frac{A\rho c^2}{3}.$$

(7.49)

Comparing Equations (7.48) and (7.49), we find that $A = 8\pi G/c^4$. As in the case of the Schwarzschild geometry, the values of unknown constants arising in the purely geometric theory of general relativity are identified with physical constants in the Newtonian theory, which were experimentally determined. So, Einstein's equations, now free of unknown constants, are

$$G^{\alpha\beta} = \frac{8\pi G}{c^4}T^{\alpha\beta}.$$

(7.50)

7.11 Energy Density and the Friedmann Equation

Having finally evaluated A, we return to the FRW model for cosmology and substitute the value of A back into Equation (7.46). The resultant equation is

$$\left(\frac{\dot{a}}{a}\right)^2 = \frac{8\pi G\rho}{3c^2} - \frac{kc^2}{a^2}.$$

(7.51)

This is known as the Friedmann equation. It is an inhomogeneous differential equation for the time evolution of the scale factor $a(t)$ in a Friedmann–Lemaître–Robertson–Walker cosmology in response to the energy density of its contents ρ for open ($k = -1$), closed ($k = +1$), and flat ($k = 0$) cosmologies.

7.12 Equations for Pressure

The remaining non-zero components of the Einstein tensor are G^{11}, G^{22}, and G^{33}. Einstein's equations (7.50) state that the resulting coefficients are proportional to the corresponding components of the stress energy tensor, so that $G^{11} = 8\pi G T^{11}/c^4$, etc. If this leads to plausible components of the stress energy tensor, then the FRW metric represents a solution of Einstein's equations. Recall from Chapter 3 that the (no sum implied) $T^{jj} = P^j$ for $j = 1, 2, 3$ and P^j is one of the three orthogonal components of the pressure. Einstein's equations for the FRW cosmology are therefore

$$G^{rr} = -\frac{(1 - kr^2)}{c^2a^2}\left(2\frac{\ddot{a}}{a} + \left(\frac{\dot{a}}{a}\right)^2 + \frac{kc^2}{a^2}\right) = \frac{8\pi G}{c^4}P^r$$

$$G^{\theta\theta} = -\frac{1}{a^2 r^2 c^2}\left(2\frac{\ddot{a}}{a} + \left(\frac{\dot{a}}{a}\right)^2 + \frac{kc^2}{a^2}\right) = \frac{8\pi G}{c^4} P^\theta$$

$$G^{\phi\phi} = -\frac{1}{a^2 r^2 c^2 \sin^2\theta}\left(2\frac{\ddot{a}}{a} + \left(\frac{\dot{a}}{a}\right)^2 + \frac{kc^2}{a^2}\right) = \frac{8\pi G}{c^4} P^\phi. \tag{7.52}$$

These are equations for the three components of pressure. Do they make sense? The dimensions are right, as you can check by substituting the correct units for c, G, and P, remembering that a and k are dimensionless, and every time derivative of a introduces a dimension of inverse time. All three pressure components have a common factor in large parentheses that arises from the curvature of the space-time (k) and the evolution of the cosmic scale (a). The other factors make each of the three spatial non-zero components of the Einstein tensor dependent on position. This at first seems at odds with the cosmological principle – because these components are proportional to the components of the pressure, shouldn't they be the same at every space-time point?

The resolution of this apparent paradox turns out to be the choice of coordinates implicit in writing down the Friedmann–Robertson–Walker metric signature of Equation (7.1). As with the Schwarzschild geometry, it turns out that the coordinate system we have chosen is a slightly peculiar one from the point of view of demanding the agreement of the resulting quantities with our expectations derived from Physics experiments.

7.13 Comoving Versus Laboratory Coordinates

We examine the coordinate system implicit in the FRW metric by considering the 11 or rr component, although exactly the same argument will follow for the 22 and 33 components. We start by recalling that $g_{\mu\nu} = \vec{e}_\mu \cdot \vec{e}_\nu$, so in the case of the rr component, we have $g_{rr} = \vec{e}_r \cdot \vec{e}_r$, so that the length of the $\vec{e}_r$ vector is $|\vec{e}_r| = \sqrt{g_{rr}} = a/\sqrt{1 - kr^2}$. This means that the basis vectors in this coordinate system change their length to track the expansion or contraction of space. Such coordinates are called comoving, because the basis vectors have lengths that scale with the expansion of the Universe.

When we are trying to relate the components of tensors to physically intuitive quantities such as pressure, being in comoving coordinates can be unhelpful. The reason why the components of the Einstein tensor in Equations (7.52) have complicated prefactors is that they are in these comoving coordinates. We can convert the radial component G^{rr} and the other spatial components to a coordinate system where the basis vectors are of unit length. We are always free to do this locally; it is equivalent to moving to the rest frame of an observer in the lab, where the walls of the lab and the equipment therein are maintained at a fixed scale by non-gravitational forces such as electromagnetism. Indeed, this is exactly the case for real labs on Earth! Hence we will refer to these as laboratory coordinates.

To work out the conversion from comoving to laboratory coordinates, note that $G^{\mu\nu}$ are components of a second-rank tensor. Considering this tensor as being a product of two first-rank tensors, as indeed it is because $T^{\alpha\beta}$ was defined as such a product in Chapter 3, we write $G^{rr} = A^r B^r$. Each of A^r and B^r are components of a vector expanded in terms of basis vectors,

$$\vec{A} = A^t \vec{e}_t + A^r \vec{e}_r + A^\theta \vec{e}_\theta + A^\phi \vec{e}_\phi$$
$$\vec{B} = B^t \vec{e}_t + B^r \vec{e}_r + B^\theta \vec{e}_\theta + B^\phi \vec{e}_\phi. \tag{7.53}$$

We can also write the same vectors in terms of new primed laboratory coordinates:

$$\vec{A} = A^{t'} \vec{e}_{t'} + A^{r'} \vec{e}_{r'} + A^{\theta'} \vec{e}_{\theta'} + A^{\phi'} \vec{e}_{\phi'}$$
$$\vec{B} = B^{t'} \vec{e}_{t'} + B^{r'} \vec{e}_{r'} + B^{\theta'} \vec{e}_{\theta'} + B^{\phi'} \vec{e}_{\phi'}. \tag{7.54}$$

We focus on the r components. Since the primed and unprimed basis vectors for each of the four directions are parallel to each other and differ only in their lengths, we can write

$$A^r \vec{e}_r = A^{r'} \vec{e}_{r'}, \tag{7.55}$$

where to convert from the comoving unprimed basis vector to laboratory basis vector, we divide by the instantaneous length of the comoving basis vector, so that

$$\vec{e}_{r'} = \frac{\vec{e}_r}{|\vec{e}_r|} = \frac{\vec{e}_r}{\sqrt{g_{rr}}}. \tag{7.56}$$

Substituting this into Equation (7.55), we obtain

$$A^r \vec{e}_r = A^r \sqrt{g_{rr}} \vec{e}_{r'}, \tag{7.57}$$

so that

$$A^{r'} = A^r \sqrt{g_{rr}}. \tag{7.58}$$

We make the same argument with the component B^r, so that the conversion for G^{rr} into our primed laboratory coordinate system is

$$G^{r'r'} = A^{r'} B^{r'} = |g_{rr}| A^r B^r = \frac{a^2}{1 - kr^2} G^{rr}. \tag{7.59}$$

When we substitute Equation (7.59) into the radial part of Equations (7.52), the factor of $|g_{rr}|^2$ cancels the position-dependent prefactor, so that

$$G^{r'r'} = -\frac{1}{c^2}\left(2\frac{\ddot{a}}{a} + \left(\frac{\dot{a}}{a}\right)^2 + \frac{kc^2}{a^2}\right) = \frac{8\pi G}{c^4} P^{r'}. \tag{7.60}$$

Exactly the same scaling argument can be made to convert the other diagonal components of $G^{\mu\nu}$ to the laboratory coordinate system. The conversion has no effect on the tt component because $|\vec{e}_t| = 1$, and hence the energy density ρ is the same in the comoving and non-comoving coordinates. This is how we were able to derive a position-independent Friedmann equation in the comoving coordinate system; the Friedmann equation is valid in either comoving or laboratory coordinates. For the $\theta\theta$ and $\phi\phi$ components, the effect is the same as that for the radial component; it removes the prefactor before the common scaling and curvature-dependent portion. We conclude that the three orthogonal components of the momentum flux or, equivalently, the pressure are equal to each other at all points in space-time, which again seems sensible given the cosmological principle of homogeneity and its companion assumption of isotropy, meaning that there is no preferred direction and no preferred position. Since all three components of the pressure are the same, we can use a single symbol P to denote all of them and write

$$2\frac{\ddot{a}}{a} + \left(\frac{\dot{a}}{a}\right)^2 + \frac{kc^2}{a^2} = -\frac{8\pi G P}{c^2}. \tag{7.61}$$

This is sometimes called the Raychaudhuri equation.

As is demonstrated through Problem 7.1 that the Raychaudhuri equation can be combined with the Friedmann equation, repeated here for reference,

$$\left(\frac{\dot{a}}{a}\right)^2 = \frac{8\pi G\rho}{3c^2} - \frac{kc^2}{a^2}, \tag{7.62}$$

to yield the following simplified equation for $\ddot{a}$, commonly called the acceleration equation,

$$\frac{\ddot{a}}{a} = -\frac{4\pi G}{3c^2}(\rho + 3P), \tag{7.63}$$

and as is shown further in Problem 7.2, by differentiating the Friedmann equation and substituting in the acceleration equation, we may obtain a third equation commonly called the fluid equation,

$$\frac{d\rho}{dt} + 3\frac{\dot{a}}{a}(\rho + P) = 0. \tag{7.64}$$

We will show in Problem 7.3 that Equation (7.64) expresses the first law of thermodynamics. It is not perhaps so surprising that the first law of thermodynamics should drop out; it is simply a statement of local energy conservation. Local energy conservation is expressed also in the divergence-free nature of the stress–energy–momentum tensor, and after all, we have seen that this very property is why Einstein selected this tensor for his equations.

7.14 The Cosmological Constant

No exploration of general relativity applied to cosmology would be complete without mention of the cosmological constant. The pathway to derivation of Einstein's equations was, as we saw, one of guesswork based on the symmetry properties of the tensors: A tensor $T^{\mu\nu}$ representing matter and energy was found to be divergence free, so that $D_\mu T^{\mu\nu} = 0$, as was another tensor $G^{\mu\nu}$ having the same rank and representing the curvature of space-time. Therefore they were deemed to be related by Einstein, making a link between the geometry of space-time and the matter and energy it contains. Einstein realised that other divergence-free tensors of the same rank could also appear in his equations. There is indeed another term having the same divergence-free characteristic. The version of Einstein's equations below has this further term included on the right:

$$G^{\alpha\beta} = \frac{8\pi G}{c^4} T^{\alpha\beta} - \Lambda g^{\alpha\beta}, \qquad (7.65)$$

where Λ is a new constant, the famous cosmological constant. The multiplicative factor $g^{\alpha\beta}$ ensures that the new term transforms correctly under changes of coordinate system – for example, the change from comoving to fixed basis vector coordinates discussed in Section 7.13. However, the term could also have been written on the left-hand side, in which case its origin could be interpreted as a correction to the geometry of the Universe.

Einstein added this term because of his personal bias towards a static universe. He subsequently removed the term when observations by Hubble revealed that the Universe is expanding, famously calling it the greatest mistake of his life. In modern cosmology and astrophysics, however, this term has taken on a new significance. Observations of the Universe on very large scales indicate that the expansion rate of the Universe is accelerating, which is at odds with expectations for a Universe whose longest range forces are dominated by the always attractive influence of the gravitational force. It turns out that the so-called cosmological constant term can be tuned so that the standard Friedmann–Robertson–Walker cosmology, including this term, matches observations of an accelerating expansion rate.

Another name sometimes given for the influence of this term is 'dark energy'. The origin of this name can be worked out by doing Problem 7.6. There we show that if the Universe only consisted of material behaving in a manner suggested by the cosmological constant term, then as the Universe expands, the energy content of the Universe rises. In some sense, the more vacuum there is, the more energy there is in the Universe!

So far, however, no acceptable physical explanation for such a term that can produce an effect of the right magnitude to match these observations is forthcoming. So, the cosmological constant term remains a purely mathematical device for parameterising an observed effect, rather than an actual explanation for the observed acceleration of the expansion of our Universe. Naturally, attempts to find a plausible physical mechanism to justify the inclusion of the cosmological constant term with a value of Λ to match the observed acceleration, and efforts to gain more data on the details of the dynamics of the expanding Universe are some of the most active research areas in modern astrophysics and cosmology.

7.15 Review

This chapter has mainly been the general relativistic derivation of the Friedmann equation, which, it is thought, governs the dynamics of the Universe, and the accompanying equations governing the pressure, or momentum flux, together with an understanding of the coordinate systems in which these equations are valid were the aim of this chapter. Note that the method by which we proceeded was exactly the same as the method used to explore the Schwarzschild geometry. Though the mathematics is laborious, it is not impossibly so, and for the author, the pedagogical value in a uniform treatment of two disparate problems and in particular the way in which we can arrive at the values of all the constants by comparison with Newton's earlier work is particularly valuable to the student and played an important role in the development of the subject by its founders.

A further discussion of cosmology and the large-scale structure of the Universe are beyond the scope of this text. Cosmology is large, and certainly the subject for another course. For some, it is the study of a lifetime. There are many very good cosmology books, notably a fine one for beginners by Ryden (2003). More advanced texts with different foci have been written by Kolb and Turner (1990), Peebles (1994), and Rubakov and Gorbunov (2017; 2011), though all of these books require a thorough grounding in astrophysics and particle physics.

7.16 Problems

7.1 By combining the Friedmann equation (7.62) with the Raychaudhuri equation (7.61), show how you derive the acceleration equation (7.63).

7.2 By differentiating a simple rearrangement of the Friedmann equation and combining this with the acceleration equation (7.63), show that you can obtain the fluid equation (7.64).

7.3 Show that the fluid equation boils down to thermodynamics. To do this, we start by writing out the first law $dU = dQ + dW$, where dU is the increase in internal energy of a system, dQ is the heat flow into the system, and dW is the work done on the system. In an isotropic homogeneous universe, there is no net heat flow, so we set $dQ = 0$. We model our Universe as a gas, so that

$dW = -P \, dV$, and the internal energy is taken to be $U = \rho V$. Substitute for dW and dU in the first law, differentiate both sides with respect to time, and show that you recover the fluid equation (7.64).

7.4 Work out the energy density and pressure corresponding to the cosmological constant term, in terms of Λ, and fundamental constants.

7.5 The equation of state of a substance can be written as a relationship between the pressure P and the energy density ρ. In general, $P = w\rho$, where w is a number. Starting from the fluid equation (7.64), show that the energy density ρ is related to the scale factor a by

$$\rho = \rho_0 a^{-3(1+w)}, \qquad (7.66)$$

where ρ_0 is the value of the scale factor when $a = 1$. We will adopt the usual convention where $a = 1$ corresponds to the present.

7.6 In kinetic theory, the magnitude P of the pressure on the walls of a gas vessel is related to the number density n of the gas particles, each mass m, and the root mean square velocity $\overline{v^2}$ of the gas particles by

$$P = \frac{1}{3} nm\overline{v^2}.$$

Considering this result, work out the equation of state of:

(a) A gas of photons;
(b) The 'dust' discussed in Chapter 3;
(c) A gas at a temperature of 2.7 K; and
(d) The cosmological constant.

In each case, solve Equation (7.66) for the dependence of the energy density ρ on the scale factor. For the particular case of a cosmological constant, show that the energy per unit volume is constant, rather as if the vacuum itself were the source of the energy. An idea from quantum field theory that the zero point energy associated with all the available modes of the vacuum might be the source of dark energy does not currently stand up to experimental scrutiny, since it produces a prediction for the abundance of vacuum energy that is far too large.

7.7 By using the Friedmann equation with $k = 0$ show that a flat Universe dominated by a cosmological constant expands exponentially.

7.8 The metric signature for a flat expanding Universe in comoving coordinates is

$$ds^2 = -c^2 \, dt^2 + a(t)^2 \left(dx_c^2 + dy_c^2 + dz_c^2 \right), \qquad (7.67)$$

where $a(t)$ is a scale factor, t is time, and x_c, y_c, and z_c are the comoving spatial coordinates. An example of such a Universe has $a(t) = \exp(Mt)$, where M is a constant. A bullet fired from a gun in this Universe has initial proper velocity v_0. The bullet travels in the direction of increasing x. There are no external forces on the bullet after it is fired. The time coordinate t is common to both proper and comoving coordinates.

(a) If the bullet is at the origin at time $t = 0$, write down an equation for $x_p(t)$, the proper coordinate position of the bullet at time t, in terms of the parameters defined above.

(b) Write equations for $\vec{X}(t)$, the vector from the origin to the bullet at time t, firstly in terms of the parameters v_0 and t, and the unit vector $\hat{e}_x$ in the direction of increasing x, and secondly in terms of the coordinate $x_c(t)$ of the bullet and the basis vector $\vec{e}_c^x$ in the comoving coordinate system.

(c) Find an equation for $x_c(t)$ in terms of v_0, t, and M.

(d) Sketch $x_c(t)$ as a function of t.

(e) At what time after the gun is fired is the comoving velocity of the bullet exactly zero? How does this time depend on the initial velocity of the bullet?

7.9   Work out the expansion of a flat Universe dominated by light as a function of time.

7.10 Work out the expansion of a flat Universe dominated by non-relativistic matter as a function of time.

8

Gravitational Waves

8.1 Astronomy with Gravity

In Chapter 7, we derived the Friedmann–Robertson–Walker metric for the Universe at large. Our understanding of cosmology is based largely on observations of light and other electromagnetic waves by astronomers. These observations are possible because electromagnetic waves travel with negligible dispersion and scattering, giving us information about the Universe from a large range of times in its history and over large distances. After all, the further an electromagnetic wave has travelled, the older the source of the wave was when it was emitted. The theory of electromagnetic radiation is based on solving Maxwell's equations for the electric and magnetic fields in a vacuum. When we apply the technique of linearisation to Einstein's equations, we will discover that they too possess wave solutions. In this chapter, we will show how gravitational waves are predicted by Einstein's equations of general relativity.

Much like electromagnetic waves, gravitational waves can travel with negligible dispersion across the Universe. In 2015 the LIGO science collaboration made the first detection of gravitational waves from an astrophysical source, two black holes colliding with each other (Abbott et al. 2016). To date, the gravitational around 100 similar black hole binary collisions have been detected by LIGO (Aasi et al. 2015) and by VIRGO (Accadia et al. 2012), another gravitational wave detection instrument. The new field of gravitational wave astronomy is particularly exciting because it offers a window on the Universe of objects which are dark – do not emit or absorb light. The Universe may look very different in gravitational waves than it does in electromagnetic ones! Gravitational astronomy will not replace electromagnetic astronomy because the detection of gravitational waves is very difficult, but even a small amount of information arising from the detection of gravitational waves is a bulwark against selection effects from only observing the Universe through its 'light' and provides powerful information towards our better understanding of the Universe.

8.2 Einstein's Equations for a Flat Vacuum

We seek wavelike solutions of Einstein's equations. Following what we know about electromagnetic waves, such waves might well take the form of fields, in this case fluctuations in the metric coefficient about an ambient background. We take this ambient background to be one with zero stress–energy content. So our waves are solutions to the vacuum Einstein's equations first encountered in Chapter 6, $G^{\alpha\beta} = 0$. We also assume that the background space-time is static and has no global curvature. From Chapter 7 we know that the Ricci curvature scalar R is zero for such a space-time. However,

$$G^{\alpha\beta} = R^{\alpha\beta} - \frac{1}{2}g^{\alpha\beta}R = 0, \tag{8.1}$$

so that

$$R^{\alpha\beta} = 0, \tag{8.2}$$

and hence

$$R_{\mu\nu} = g_{\mu\alpha}g_{\nu\beta}R^{\alpha\beta} = 0. \tag{8.3}$$

We can express these flat space Einstein equations in terms of the Christoffel symbols using the definition of the Riemann curvature:

$$R^{\alpha}{}_{\beta\gamma\delta} = \frac{\partial \Gamma^{\alpha}_{\delta\beta}}{\partial x^{\gamma}} - \frac{\partial \Gamma^{\alpha}_{\gamma\beta}}{\partial x^{\delta}} + \Gamma^{\alpha}_{\gamma\kappa}\Gamma^{\kappa}_{\delta\beta} - \Gamma^{\alpha}_{\delta\kappa}\Gamma^{\kappa}_{\gamma\beta}, \tag{8.4}$$

so that the flat space-time vacuum Einstein equations become

$$R_{\beta\delta} = R^{\alpha}{}_{\beta\alpha\delta} = \frac{\partial \Gamma^{\alpha}_{\delta\beta}}{\partial x^{\alpha}} - \frac{\partial \Gamma^{\alpha}_{\alpha\beta}}{\partial x^{\delta}} + \Gamma^{\alpha}_{\alpha\kappa}\Gamma^{\kappa}_{\delta\beta} - \Gamma^{\alpha}_{\delta\kappa}\Gamma^{\kappa}_{\alpha\beta} = 0. \tag{8.5}$$

8.3 Linear Metric Perturbations

Einstein's equations are nonlinear: when written out in full, the metric coefficients and their derivatives appear to many powers and in many combinations. However, if we assume that the size of the perturbations about the ambient flat vacuum background is very small, then any terms that are nonlinear in these perturbations or their derivatives can be neglected. This process of assuming small perturbations about some understood operating point and neglecting nonlinear terms in these perturbations is called linearisation. It is a powerful and common technique in handling complex differential equations of all kinds. It is required that you decide upon an operating point at which the equations in question have well understood behaviour; for us, that operating point is flat and empty space-time. To linearise

Einstein's equations, we write the metric in Cartesian coordinates and assume that our disturbances are small perturbations on the flat space-time metric signature:

$$g_{\mu\nu}(x) = \eta_{\mu\nu} + h_{\mu\nu}(x), \tag{8.6}$$

where $\eta_{\mu\nu} = \text{diag}(-1, 1, 1, 1)$, and the coefficients $h_{\mu\nu}$ are functions of position and time but always have magnitudes very much less than 1.

To substitute these metric coefficients into Einstein's equations, we will need to evaluate the Christoffel symbols. We do this using Equation (4.79), obtaining

$$\Gamma^{\alpha}_{\beta\gamma} = \frac{(\eta^{\alpha\omega} + h^{\alpha\omega})}{2} \left(\frac{\partial h_{\omega\beta}}{\partial x^{\gamma}} + \frac{\partial h_{\omega\gamma}}{\partial x^{\beta}} - \frac{\partial h_{\beta\gamma}}{\partial x^{\omega}} \right), \tag{8.7}$$

where we have left out the partial derivatives of $\eta_{\mu\nu}$ because they are zero. However, the terms where $h^{\alpha\omega}$ multiplies the derivatives of h are nonlinear in h and are therefore negligible. The Christoffel symbols become

$$\Gamma^{\alpha}_{\beta\gamma} = \frac{\eta^{\alpha\omega}}{2} \left(\frac{\partial h_{\omega\beta}}{\partial x^{\gamma}} + \frac{\partial h_{\omega\gamma}}{\partial x^{\beta}} - \frac{\partial h_{\beta\gamma}}{\partial x^{\omega}} \right). \tag{8.8}$$

Before we substitute these Christoffel symbols into the vacuum flat Einstein's Equations (8.5), we can make one further simplification. The terms that contain a product of two Christoffel symbols are always at least quadratic in h, and hence these terms are also negligible. So, the linearised vacuum Einstein equations become

$$\frac{\partial \Gamma^{\alpha}_{\delta\beta}}{\partial x^{\alpha}} - \frac{\partial \Gamma^{\alpha}_{\alpha\beta}}{\partial x^{\delta}} = 0. \tag{8.9}$$

Substituting Equation (8.8) into Equation (8.9), we obtain

$$\frac{\partial \Gamma^{\alpha}_{\delta\beta}}{\partial x^{\alpha}} - \frac{\partial \Gamma^{\alpha}_{\alpha\beta}}{\partial x^{\delta}} = \frac{\eta^{\alpha\omega}}{2} \left(\frac{\partial^2 h_{\omega\delta}}{\partial x^{\alpha}\partial x^{\beta}} + \frac{\partial^2 h_{\omega\beta}}{\partial x^{\alpha}\partial x^{\delta}} - \frac{\partial^2 h_{\delta\beta}}{\partial x^{\alpha}\partial x^{\omega}} \right.$$
$$\left. - \frac{\partial^2 h_{\omega\alpha}}{\partial x^{\delta}\partial x^{\beta}} - \frac{\partial^2 h_{\omega\beta}}{\partial x^{\delta}\partial x^{\alpha}} + \frac{\partial^2 h_{\alpha\beta}}{\partial x^{\delta}\partial x^{\omega}} \right)$$
$$= \frac{\eta^{\alpha\omega}}{2} \left(\frac{\partial^2 h_{\omega\delta}}{\partial x^{\alpha}\partial x^{\beta}} - \frac{\partial^2 h_{\delta\beta}}{\partial x^{\alpha}\partial x^{\omega}} - \frac{\partial^2 h_{\omega\alpha}}{\partial x^{\delta}\partial x^{\beta}} + \frac{\partial^2 h_{\alpha\beta}}{\partial x^{\delta}\partial x^{\omega}} \right) = 0. \tag{8.10}$$

The second term is

$$-\frac{1}{2} \eta^{\alpha\omega} \frac{\partial}{\partial x^{\alpha}} \frac{\partial}{\partial x^{\omega}} h_{\delta\beta} = \frac{1}{2} \left(\frac{1}{c^2} \frac{\partial^2}{\partial t^2} - \nabla^2 \right) h_{\delta\beta}. \tag{8.11}$$

Notice that if this term is set equal to zero, this is then a wave equation for the components $h_{\delta\beta}$,

$$\left(\frac{1}{c^2} \frac{\partial^2}{\partial t^2} - \nabla^2 \right) h_{\delta\beta} = 0. \tag{8.12}$$

So, here is how we will proceed. We will assert that there is a wave solution by setting this term equal to zero. If this is right, then we will have wave solutions for the metric perturbations, $h_{\delta\beta}$:

$$h_{\delta\beta}(ct, \vec{r}) = a_{\delta\beta} \cos(\vec{k} \cdot \vec{r} - \omega t + \phi_0), \tag{8.13}$$

where $a_{\delta\beta}$ are the amplitudes of the waves in the different components of the metric, $\vec{k}$ is the wave vector, ω is the angular frequency, and ϕ_0 is a phase offset. Without loss of generality, we can choose the wave vector $\vec{k}$ to be in the z direction, so that

$$h_{\delta\beta}(ct, z) = a_{\delta\beta} \cos(kz - \omega t + \phi_0). \tag{8.14}$$

For all this to work, the other three terms in Equation (8.10) have to sum to zero as well:

$$\frac{\partial^2 h_\delta^\alpha}{\partial x^\beta \partial x^\alpha} + \frac{\partial^2 h_\beta^\alpha}{\partial x^\delta \partial x^\alpha} - \frac{\partial^2 h_\alpha^\alpha}{\partial x^\beta \partial x^\delta} = 0. \tag{8.15}$$

Note however that the wave solutions of Equation (8.14) do not have any x or y dependence. Therefore any partial derivative of the wave with respect to x or y is zero. So, if $\beta = 1$ or $\beta = 2$, then only the middle term on the left survives. Furthermore, even in this remaining term, those where $\alpha = 1$ or $\alpha = 2$ vanish for the same reason. So, we list the remaining terms for $\beta = 1$ and then for $\beta = 2$:

$$\frac{\partial^2 h_1^0}{\partial x^\delta \partial x^0} + \frac{\partial^2 h_1^3}{\partial x^\delta \partial x^3} = 0$$

$$\frac{\partial^2 h_2^0}{\partial x^\delta \partial x^0} + \frac{\partial^2 h_2^3}{\partial x^\delta \partial x^3} = 0. \tag{8.16}$$

However, for the same reason, we cannot have $\delta = 1$ or $\delta = 2$ in Equations (8.16). So we are left with

$$\frac{\partial^2 h_1^0}{\partial (x^0)^2} + \frac{\partial^2 h_1^3}{\partial x^0 \partial x^3} = 0$$

$$\frac{\partial^2 h_2^0}{\partial x^3 \partial x^0} + \frac{\partial^2 h_2^3}{\partial (x^3)^2} = 0. \tag{8.17}$$

One way these second derivatives can add to zero is if h_1^0, h_1^3, h_2^0, and h_2^3 are zero. Furthermore, because $h_{\delta\beta}$ is symmetric, this implies that h_0^1, h_3^1, h_0^2, and h_3^2 are all zero as well.

Now we substitute these zero terms into Equation (8.15). If we set $\beta = \delta = 0$, then we obtain

$$\frac{\partial^2 h_0^0}{\partial (x^0)^2} + 2\frac{\partial^2 h_0^3}{\partial x^0 \partial x^3} - \frac{\partial^2}{\partial (x^0)^2}(h_1^1 + h_2^2) - \frac{\partial^2 h_3^3}{\partial (x^0)^2} = 0. \tag{8.18}$$

If, instead, we set $\beta = 3$ and $\delta = 0$, then we obtain

$$\frac{\partial^2 h_0^3}{\partial (x^3)^2} - \frac{\partial^2}{\partial x^3 \partial x^0}\left(h_1^1 + h_2^2\right) - \frac{\partial^2 h_3^3}{\partial x^3 \partial x^0} = 0. \tag{8.19}$$

Finally, if we set $\beta = \delta = 3$, then we obtain

$$2\frac{\partial^2 h_3^0}{\partial x^3 \partial x^0} + \frac{\partial^2 h_3^3}{\partial (x^3)^2} - \frac{\partial^2}{\partial (x^3)^2}\left(h_0^0 + h_1^1 + h_2^2\right) = 0. \tag{8.20}$$

Equations (8.18), (8.19), and (8.20) are all satisfied if h_0^0, h_3^3, h_0^3, and h_3^0 are all zero and $h_1^1 + h_2^2 = 0$. Note that though we have worked out these relationships using rank $^1_1$ components h_α^ν, the same relationships apply to their index-lowered rank $^0_2$ cousins. This is again because we are working to first order in h_α^ν, so that we can lower the upstairs index in h_α^ν using $\eta_{\mu\nu}$.

We have therefore established that we can satisfy Equation (8.15) if the propagation direction for the waves is parallel to the z axis and if the only non-zero elements of $h_{\alpha\beta}$ are those with α and β each equal to 1 or 2. Furthermore, $h_{11} + h_{22} = 0$, and by the symmetry of $g_{\mu\nu}$, $h_{12} = h_{21}$. If these conditions are satisfied, then the wave solutions of Equation (8.14) are valid solutions of Einstein's equations in the limit where $h_{\alpha\beta} \ll 1$.

The form for the elements of $g_{\mu\nu}$ in the presence of this wave solution is therefore

$$(g_{\mu\nu}) = \begin{pmatrix} -1 & & & \\ & +1 & & \\ & & +1 & \\ & & & +1 \end{pmatrix} + \begin{pmatrix} 0 & & & \\ & h_+ & h_\times & \\ & h_\times & -h_+ & \\ & & & 0 \end{pmatrix}\cos(k_z z - \omega t + \phi_0), \tag{8.21}$$

where h_+ and $h_\times$ are two independent amplitudes of two components of the propagating wave.

8.4 Gauges in Gravitation

If you have studied advanced electromagnetism, then you will know that the electric and magnetic fields can be expressed as derivatives of a vector four-potential whose components are given the symbols A^μ. This four-potential is not an observable, and hence there is some freedom to re-define the components A^μ as long as the derivatives of the modified components still yield the same observable electric and magnetic fields. Because it can be simpler to manipulate the vector potential than the electric and magnetic fields, particular choices of vector potential can be convenient. The freedom to make certain transformations on the components of the vector potential and not affect the observable fields is known as gauge freedom. The different choices have been given names; for example, there are the Coulomb gauge and the Lorentz gauge. The names reflect the properties of the vector potential that arise from the particular choice.

In gravitation, it turns out that we can make certain choices of the properties of the components of the metric. This is also referred to as a choice of gauge. However, there is an important difference. In gravitation, the metric coefficients are observable quantities. For example, LIGO has measured the components of the metric in the vicinity of its detectors during the passage of gravitational waves through the Earth. When we make a choice of gauge in gravitation, we are picking a particular coordinate system in which an observer might measure the metric co-efficients. There is no unobservable field analogue of the vector potential. Though the language that has grown up is the same, choices of gauge in gravitation are markedly different from choices of gauge in electromagnetism.

To make a gauge transformation in general relativity, you make a transformation of coordinate system. Let x^α be the coordinate of a point in an unprimed coordinate system. Then the transformation to a primed coordinate system leads to a primed coordinate for the same point,

$$x^{\alpha'} = x^\alpha + \varepsilon^\alpha(x). \tag{8.22}$$

Here we continue to adhere to the convention that the prime lives with the in-dex; unfortunately, this leads to the same index appearing primed and unprimed on either side of this equation, but this will not cause confusion since we are about to figure out the consequences of the gauge transformation for tensors, whereupon this will not matter anymore. Differentiating with respect to x^β, we obtain

$$\frac{\partial x^{\alpha'}}{\partial x^\beta} = \frac{\partial x^\alpha}{\partial x^\beta} + \frac{\partial \varepsilon^\alpha}{\partial x^\beta}$$

$$= \delta^\alpha_\beta + \frac{\partial \varepsilon^\alpha}{\partial x^\beta}. \tag{8.23}$$

We can also differentiate Equation (8.22) with respect to $x^{\gamma'}$, obtaining

$$\frac{\partial x^{\alpha'}}{\partial x^{\gamma'}} = \frac{\partial x^\alpha}{\partial x^{\gamma'}} + \frac{\partial \varepsilon^\alpha}{\partial x^{\gamma'}}$$

$$\delta^{\alpha'}_{\gamma'} = \frac{\partial x^\alpha}{\partial x^{\gamma'}} + \frac{\partial \varepsilon^\alpha}{\partial x^{\gamma'}}$$

$$\frac{\partial x^\alpha}{\partial x^{\gamma'}} = \delta^{\alpha'}_{\gamma'} - \frac{\partial \varepsilon^\alpha}{\partial x^{\gamma'}}. \tag{8.24}$$

We assume that the components representing the shift in coordinates $\varepsilon^\lambda(x)$ are small parameters relative to the coordinates themselves. This can always be made the case by a suitable choice of origin. This being the case, having introduced a single factor of ε^α to effect the gauge transformation, we can then replace all the primed symbols on the right with unprimed ones, since any corrections between unprimed and primed symbols will involve further factors of ε^λ, which can be neglected to first order. So, we arrive at

$$\frac{\partial x^\alpha}{\partial x^{\gamma'}} = \delta^\alpha_\gamma - \frac{\partial \varepsilon^\alpha}{\partial x^\gamma}. \tag{8.25}$$

Having the two partial derivatives in Equations (8.23) and (8.25), we can work out the effect of the gauge transformation on the components of any tensor. For our purposes, the important example is the metric components, which as tensors of rank $^0_2$ transform as

$$g_{\alpha'\beta'} = \frac{\partial x^\mu}{\partial x^{\alpha'}} \frac{\partial x^\nu}{\partial x^{\beta'}} g_{\mu\nu}$$
$$= \left(\delta^\mu_\alpha - \frac{\partial \varepsilon^\mu}{\partial x^\alpha}\right)\left(\delta^\nu_\beta - \frac{\partial \varepsilon^\nu}{\partial x^\beta}\right) g_{\mu\nu}. \tag{8.26}$$

Again, to first order in ε^λ, we can neglect the term in the product of the differentials and arrive at

$$g_{\alpha'\beta'} \simeq \left(\delta^\mu_\alpha \delta^\nu_\beta - \delta^\mu_\alpha \frac{\partial \varepsilon^\nu}{\partial x^\beta} - \delta^\nu_\beta \frac{\partial \varepsilon^\mu}{\partial x^\alpha}\right) g_{\mu\nu}$$
$$= g_{\alpha\beta} - g_{\alpha\nu} \frac{\partial \varepsilon^\nu}{\partial x^\beta} - g_{\mu\beta} \frac{\partial \varepsilon^\mu}{\partial x^\alpha}. \tag{8.27}$$

This shows that gauge transformations also cause offsets in the components of the metric. As we will discuss in Section 8.5, the transverse perturbations on the metric that represent our plane gravitational waves correspond once again to the viewpoint of a rather special class of observer. The importance of the above analysis is that it shows how to relate a transformation on the metric coefficients to a specific gauge transformation on the coordinate system that gives rise to it.

8.5 Observers of the Transverse Traceless Gauge

We have discovered that to some observer the gravitational waves appear transverse – the disturbances to the metric coefficients are in directions perpendicular to the propagation direction and are traceless – the diagonal elements of the disturbance matrix sum to zero. Just as for the Schwarzschild geometry and for the Friedman–Robertson–Walker cosmology, now that we have found a solution to Einstein's equations, we should ask whose coordinates are we in here? One thing is clear – these are not lab coordinates! Every time you see metric coefficients with time dependence, you know that this is a sign of a non-laboratory frame of reference. Labs do not stretch and distort with the passage of time under the influence of gravitation!

Another sign that this coordinate system does not correspond to a lab is that for a beam propagating down the x-arm, we can write down an expression for dx/dt for the photon. From Equation (8.39) we get

$$\frac{dx}{dt} = \frac{c}{(1 + h_+ \cos(\omega t))^{1/2}}. \tag{8.28}$$

In these coordinates, the rate of change of x with respect to time, the rate at which the photons in the light beam pass the increasing values of x, oscillates with the gravitational wave. This does not of course mean that the speed of light is oscillating; it rather means that the coordinate system itself is being distorted with respect to laboratory coordinates as the gravitational wave passes through. In summary, the waves are only transverse and traceless in a coordinate system corresponding to some observers who are accelerating with respect to any reasonable 'lab frame'. To investigate this further, we follow the same prescription as in the previous chapter by finding out the lengths of the basis vectors. In the case of a gravitational wave having h_+ non-zero and $h_\times = 0$, we have

$$|\vec{e}_x| = \sqrt{g_{xx}} = \sqrt{1 + h_+ \cos(\omega t)}$$
$$|\vec{e}_y| = \sqrt{g_{yy}} = \sqrt{1 - h_+ \cos(\omega t)}. \tag{8.29}$$

Not surprisingly, we discover that these basis vectors get longer and shorter to follow the gravitational wave signal. So, in these coordinates, it turns out that the coordinate positions of the test masses do not change! The coordinate system is a mesh that stretches and distorts with the gravitational wave. For a body in free fall in this coordinate system, the coordinates x and y of the bodies position are static.

8.6 Laboratory Coordinates

Still following the treatment in Chapter 7, we define a lab coordinate system in which the lengths of the basis vectors are locally 1. We do this by considering a vector to a point in the XY plane, $\vec{r} = x\vec{e}_x + y\vec{e}_y$. We write a new lab coordinate basis vector $\vec{e}_x{}^L$ of unit length. We play the same trick with the y direction. We obtain

$$\vec{e}_x{}^L = \frac{\vec{e}_x}{|\vec{e}_x|} = \frac{\vec{e}_x}{\sqrt{g_{xx}}}$$
$$\vec{e}_y{}^L = \frac{\vec{e}_y}{|\vec{e}_y|} = \frac{\vec{e}_y}{\sqrt{g_{yy}}}. \tag{8.30}$$

Our vector can now be written

$$\vec{r} = x\vec{e}_x + y\vec{e}_y = x\sqrt{g_{xx}}\vec{e}_x{}^L + y\sqrt{g_{yy}}\vec{e}_y{}^L. \tag{8.31}$$

Because the factors of $\sqrt{g_{xx}}$ and $\sqrt{g_{yy}}$ are time dependent, the vector $\vec{r}$ is time dependent too. However, the basis vectors $\vec{e}_x{}^L$ and $\vec{e}_y{}^L$ have no time dependence; by construction they are static unit vectors. Therefore the components $x^L = x\sqrt{g_{xx}}$ and $y^L = y\sqrt{g_{yy}}$ carry the time dependence. These components can be thought of as the coordinates of a particle free to respond to the motion of the gravitational wave:

$$\vec{r}(t) = x^L(t)\vec{e}_x{}^L + y^L(t)\vec{e}_y{}^L$$

$$x^L(t) = x\left(1 + \frac{h_+}{2}\cos(\omega t)\right)$$

$$y^L(t) = y\left(1 - \frac{h_+}{2}\cos(\omega t)\right). \tag{8.32}$$

Notice that the transformations between the transverse traceless coordinate system and the lab coordinates are of the form given in Equation (8.22). This will be the subject of Problem 8.3. The size of the motions scale linearly with the distance of the point from the origin. This may seem mildly surprising – what is special about the distance from an arbitrary origin? The answer is – the origin of a local laboratory-based coordinate system has to be somewhere! This is our chosen reference point. Were we to move the origin somewhere else, we would find the same sized displacements of freely falling objects, proportional to the distance from the new origin. The thing that is special about the origin is that it is the one where we chose to place the LIGO beamsplitter, and hence the natural location of the origin of our coordinates.

Now we can work out the accelerations that appear to act on masses in free fall in the lab coordinates. Imagine a matrix of test masses suspended from wires in the XY plane. A gravitational wave passes through. Their displacements are subject to oscillations that depend on their initial coordinates (x, y). Because each of the components of their motion obeys the simple harmonic oscillator equation, the acceleration components are equal to the minus the square of the angular frequency times their displacements. Therefore the acceleration of a body at position (x, y) is given by

$$\frac{d^2}{dt^2}\begin{pmatrix} x^L(t) \\ y^L(t) \end{pmatrix} = \frac{\omega^2 h_+}{2}\begin{pmatrix} -x \\ +y \end{pmatrix}\cos(\omega t). \tag{8.33}$$

The direction of the acceleration at time $t = 0$ is

$$\theta = -\arctan\left(\frac{y}{x}\right). \tag{8.34}$$

The directions and magnitudes of the accelerations of test masses as functions of position on the XY plane are shown in Figure 8.1.

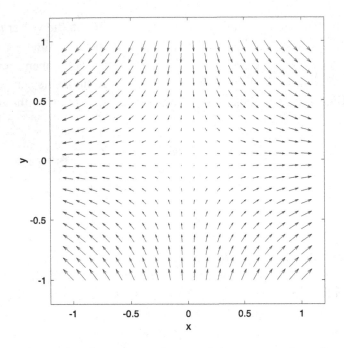

Figure 8.1 The acceleration vectors of suspended test masses in a plus polarized gravitational wave field with the waves incident into the plane in the + polarisation.

8.7 Cross and Plus Polarised Waves

The waves we have considered so far have had h_+ non-zero and $h_\times = 0$. Now we will consider another possibility, that with $h_\times$ non-zero and $h_+ = 0$. In this case the XY plane matrix coefficients in the original transverse traceless gauge coordinates are

$$\begin{pmatrix} g_{xx} & g_{xy} \\ g_{yx} & g_{yy} \end{pmatrix} = \begin{pmatrix} 0 & 1 \\ 1 & 0 \end{pmatrix} h_\times \cos(\omega t). \tag{8.35}$$

It will turn out to be interesting to consider a new coordinate system, one where the new basis vectors are defined in terms of the original ones by

$$\vec{e}_x' = \frac{\vec{e}_x + \vec{e}_y}{\sqrt{2}}$$

$$\vec{e}_y' = \frac{\vec{e}_x - \vec{e}_y}{\sqrt{2}}. \tag{8.36}$$

These new basis vectors are at right angles to each other, and the pair of primed basis vectors is inclined at 45 degrees to the pair of unprimed basis vectors. Furthermore, in Problem 8.5, you will show that in the new primed coordinate system the metric coefficients become

$$\begin{pmatrix} g_{x'x'} & g_{x'y'} \\ g_{y'x'} & g_{y'y'} \end{pmatrix} = \begin{pmatrix} 1 & 0 \\ 0 & -1 \end{pmatrix} h_\times \cos(\omega t). \tag{8.37}$$

In other words, the cross-polarised waves impose the same pattern of polarisations in the plane, only the pattern is at 45 degrees to the pattern already seen for the plus-polarised waves.

You might initially be puzzled that we are able to call these two modes different polarisations when there is an angle of forty five degrees between them. However, it turns out that when you overlay the pattern of vector accelerations of the cross and plus polarisations, you discover that at every point in the plane the acceleration of a test mass freely suspended is at right angles for the two polarisations. It is just that the direction of the force is not uniform across the plane for either of the two polarisations. In this respect, the forces on test particles in the case of gravitational waves are different from the patterns of forces on test charges in an electromagnetic wave field. The magnitudes and directions of the accelerations due to the two polarisations across the XY plane are shown in Figure 8.2. The usual way of visualising the motion of test masses in the two polarisations of the wave is to

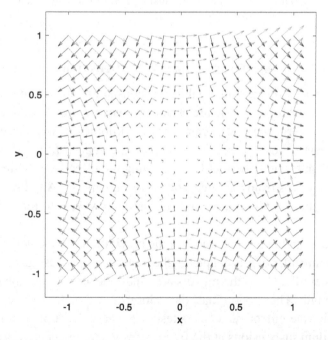

Figure 8.2 The acceleration vectors of suspended test masses in the plus (black) and cross (grey) polarized gravitational wave field with the waves incident into the plane. Notice that at all the points, the accelerations due to the two polarisations of gravitational waves are orthogonal to each other.

imagine a ring of floating test particles with plus- and cross-polarised gravitational waves incident normal to the plane of the ring. The motion of the test masses in a single period of the gravitational wave with the amplitude greatly exaggerated is given in Figure 8.3. The strain is the ratio of the amount by which the circle is distorted to the dimension of the circle.

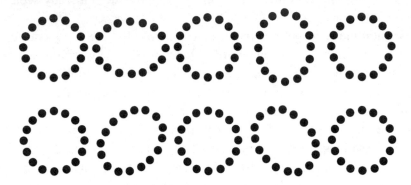

Figure 8.3 The motion of a ring of freely floating test masses, not connected to each other, during a single period of plus (upper) and cross (lower) polarised gravitational waves. The effect has been greatly exaggerated; for astrophysical sources of reasonable power, the expected signal would correspond to a distortion of about 10^{-23} of the diameter of the circle.

8.8 Gravitational Wave Detection

The LIGO detectors for gravitational waves consist of Michelson interferometers illuminated by laser light. Lasers are necessary because only a laser beam has sufficiently small divergence that it can travel for kilometers with acceptably small angular divergence. For reasons that will be come apparent, the Michelson interferometer configuration is perfect for discerning the effects on the instrument due to gravitational waves from the effects of background noise. An ultra-simplified diagram of the optical configuration of a gravitational wave interferometer is shown in Figure 8.4.

We now work out the consequences of the gravitational wave in the detector. We work for the remainder of this chapter in the transverse traceless gauge coordinates, in which the coordinates of the mirrors are time-independent but the basis vectors oscillate in length. These coordinates are closer to the reality of the experimental apparatus, since the mirrors that act as test masses are decoupled from the laboratory by pendulum suspensions at the frequencies where the detectors are sensitive. Work through Problem 8.2 to see how this works.

Let us for now only consider the horizontal arm and the passage of light from the beam splitter to the x-end mirror. Because we are using a light beam and because

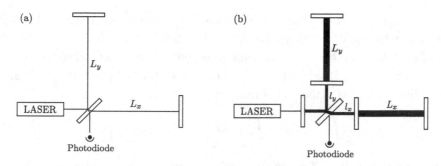

Figure 8.4 Simplified block diagrams of the a ground based interferometric grav-itational wave detector. You can see how, comparing with Figure 8.3, the end mirrors will move in antiphase if they are suspended from wires so that they can respond to an incoming plus-polarised gravitational wave. In (a), you see the sim-plest imaginable Michelson interferometer. The sensitivity of a simple Michelson interferometer to gravitational waves is discussed in Sections 8.8 and 8.9. In (b), I have added arm cavities and a recycling mirror. The utility of these modifications for enhancing the sensitivity of the instruments is discussed in Section 8.9.2.

the beam direction is aligned with our chosen x-axis, we have $ds^2 = 0$ and $dy = dz = 0$. Also, we assume that the wave is arriving directed down the z axis, so that $k_x = k_y = 0$ and $\vec{k} = k_z \vec{e_z}$, where $\vec{e_z}$ is a basis vector in the transverse traceless gauge coordinate system. We set the unknown phase offset so that $kz + \phi_0 = 0$ in Equation (8.21), and we also consider the amplitude $h_\times = 0$ for now. Consequently, for our propagating light beam, we have

$$c^2 \, dt^2 = dx^2 \left(1 + h_+ \cos(\omega t)\right). \tag{8.38}$$

This is a highly idealised set of assumptions. The detector groups have to grapple with what happens when the wave arrives from an arbitrary sky location and has some mixture of h_+ and $h_\times$ components, as well as many other so-called 'nuisance' parameters – parameters of the experiment that are called a nuisance because we have no control over them. For clarity, we disregard all of these effects and just assume that for this wave, we got really lucky, and everything is perfectly set up for a successful detection. Because the laser beam is travelling in the direction of increasing x, we take the square root of Equation (8.38) and arrive at

$$dx = \frac{c \, dt}{(1 + h_+ \cos(\omega t))^{1/2}}. \tag{8.39}$$

Carrying on with our tradition of linearising everything we can, we note that $h_+ \ll 1$, and, consequently, we write

$$dx = c \, dt \left(1 - \frac{h_+}{2} \cos(\omega t)\right), \tag{8.40}$$

where the square root in the denominator has been turned into a factor of $-1/2$ in the brackets by use of a binomial expansion to first order in h_+. This is the same kind of trick we have been playing throughout our discussion of linearised gravity. We now integrate both sides along the trajectory of the light as it travels from the beamsplitter at $x = 0$ to the end of the arm at $x = L_x$:

$$\int_{x=0}^{L_x} dx = c \int_{t=0}^{T} dt \left(1 - \frac{h_+}{2} \cos(\omega t) \right). \tag{8.41}$$

Though we can easily perform the integrals as they are in this linear approximation, we choose to take advantage of a further practical simplification. The gravitational waves we are sensitive to in LIGO have frequencies of order 100 Hz. This means that their wavelengths are around c/v, which is about 3 million metres. Consequently, in the time the light takes to travel down the arm and back again, the change of phase of the gravitational wave signal is negligible. In practice, we can factor out the term containing the $\cos(\omega t)$ and regard this term as static over the travel time of the light. You might think that this simplification is a step too far, but in fact there is still enough physics left in the equations to get to the useful result. The equation becomes very simple indeed:

$$\int_{x=0}^{L_x} dx = c \left(1 - \frac{h_+}{2} \cos(\omega t) \right) \int_{t=0}^{T} dt, \tag{8.42}$$

so that

$$L_x = cT \left(1 - \frac{h_+}{2} \cos(\omega t) \right). \tag{8.43}$$

The quantity that will be measurable, as we will see, is the time of flight. However, the time of flight in one direction from the beamsplitter to the end mirror is not in fact measurable. What is measurable is the round trip time of flight, which we shall call T_x, equal to $2T$, and given by

$$T_x = \frac{2L_x}{c(1 - h_+/2 \cos(\omega t))}. \tag{8.44}$$

Using the linear approximation again, we employ a first-order binomial expansion to simplify this to

$$T_x = \frac{2L_x}{c} \left(1 + \frac{h_+}{2} \cos(\omega t) \right)$$

$$= \frac{2L_x}{c} + \frac{L_x h_+}{c} \cos(\omega t). \tag{8.45}$$

Therefore the time of flight of photons down to the end of the x-arm and back to the beamsplitter is observably affected by the passage of the gravitational wave at frequency ω. Why, then, not just build a gravitational wave detector consisting of a single arm and search for fluctuations in the light propagation time down that arm? In fact, gravitational waves at very low frequencies have been searched for that way using the corner cube retroreflecting mirror that one of the early moon landing astronauts placed on the lunar surface. By firing laser pulse trains at that mirror and recording the times of flight you can look for very-low-frequency gravitational waves! The experiments have been done. They did not detect anything because they are not very sensitive for various technical reasons that we do not have time to explore here.

Why not repeat the same kind of trick for a ground-based detector? The main reason is that there are sources of noise that cause fluctuations in the other quantities that T_x depends on. You can essentially consider any of the symbols in Equation (8.45) to be subject to noise fluctuations. For example, you can see L_x, which is the difference between the positions of the beamsplitter and the X-end mirror. For the above analysis to work, the mirrors and beamsplitter must be on geodesics, that is, they must be in free fall in the gravitational field. In fact, they are suspended in vacuum from fine fused silica fibres. They are, effectively, the masses at the end of pendulums. Superficially, these are not then in free fall. However, in terms of the component of their motion horizontally, in the XY plane, the mirrors are almost free to move at frequencies well above the natural frequency of the pendulum, which for LIGO is about 1 Hz. This will be the subject of Problem 8.2. Consequently, for frequencies around 100 Hz, where LIGO is trying to detect gravitational waves, the mirrors are effectively free to move. This means that other sources of disturbance can, and do, move the mirrors, and by distances far larger than the amplitudes of the gravitational wave signals we are trying to detect. For example, the displacement of a test mass due to seismic noise can easily be of order a micrometre. The equivalent displacement of the mirror surface when a gravitational wave is incident is less than 10^{-18} m, which is a thousandth of a proton diameter.

The elimination of technical noise by something like 12 orders of magnitude to enable detection of gravitational waves is the real story of LIGO! From first concept to successful first detection took 40 years. The start of this story is the classic article by the Weiss, which is a lesson in how to draft a really good experiment proposal (Weiss 1972). Suffice to say that those 40 years were spent in large part working out and eliminating through careful design all of those technical noise sources, many of which were anticipated by Weiss. The book by Saulson (2017) and the references therein are an excellent springboard into an understanding of the technical issues and detector technology. The author is privileged to have played a small part in the effort, to have had the company of many of the worlds finest physicists in this work and to have learned much from them. Though this is a book about gravity, I have taken the liberty in the ensuing sections of discussing several pieces of non-gravitational physics that are important to the design and operation of gravitational wave detectors like LIGO.

8.9 Differential Interferometry

LIGO uses two perpendicular arms at right angles in the Michelson interferometer configuration. The reason for this is that the form of the signal exactly matches the degree of freedom that the Michelson interferometer is sensitive to. When the gravitational wave arrives, one arm will get longer while the other is simultaneously getting shorter. By contrast, many of the technical noise sources that afflict LIGO mostly perturb the average arm length $(L_x + L_y)/2$. For example, if the laser wavelength shifts, then this simultaneously shifts the number of waves that will fit into the two arms in the same direction. The use of the Michelson interferometer separates many of the technical noise sources, which excite the common mode from the gravitational wave signal, which excites the differential length defined as $(L_x - L_y)/2$. Of course, the experimental reality is not that simple; instrumentalists talk about the 'common mode rejection ratio', meaning how efficiently common mode noise sources are actually rejected in the differential degree of freedom.

The analysis of the passage of light down the y-arm proceeds exactly the same as that for the x-arm, only Equation (8.21) shows that there is a relative minus sign in the disturbance between the g_{xx} and g_{yy} metric coefficients. This minus sign propagates through the preceding analysis, leading to a slightly different expression for T_y, the round trip time of flight down the y-arm:

$$T_x = \frac{2L_x}{c} + \frac{L_x h_+}{c}\cos(\omega t)$$

$$T_y = \frac{2L_y}{c}\left(1 - \frac{h_+}{2}\cos(\omega t)\right)$$

$$= \frac{2L_y}{c} - \frac{L_y h_+}{c}\cos(\omega t), \tag{8.46}$$

where I have reproduced Equation (8.38), so you can see that the difference between T_x and T_y is linear in the amplitude h_+:

$$T_x - T_y = \frac{2(L_x - L_y)}{c} + \frac{(L_x + L_y)h_+}{c}\cos(\omega t). \tag{8.47}$$

We still have to consider how to detect the difference in times of flight. In reality, the two arms are not independent, but are components of the Michelson interferometer. Light exits the laser and is split into two components, which propagate down the two arms and return to the beamsplitter. Let us make a simple analysis of this physics using plane wave optics. This is again a huge simplification of the detectors, where laser beams are not at all like plane waves! Again, it is a sophisticated enough model that we can gain understanding from it. This is what physics models are all about. If we consider a beam of wavenumber $k = 2\pi v/c$, then the phase shift that the beam acquires in travelling back and forth down the x (y) arm is $2kL_x$ ($2kL_y$). Furthermore, when the two beams were formed at the beamsplitter, the y beam bounces off the beamsplitter near surface to enter the arm, but the x beam must transmit through the beamsplitter twice, once on incidence into the arm and once on return, before returning to the same point. As a consequence, the x beam acquires a minus sign in its amplitude relative to the y beam. The origin of this minus sign is subtle. It originates in the invariance of electromagnetic fields under time reversal.

The two beams returning to the beamsplitter recombine coherently, so that you can represent their amplitudes using the usual complex representation for optical phases with the correct inserted values of the phase shifts in the arms. We take T_x to be the round trip time taken for the beam to traverse the x-arm in both directions, as before. The optical field of the recombining beams Ψ, the amplitude of the beam at recombination, is

$$\Psi = \frac{1}{2}A_0 e^{i\omega_c T_x} - \frac{1}{2}A_0 e^{i\omega_c T_y}, \tag{8.48}$$

where A_0 is the amplitude of the incident laser light in each beam, and ω_c is the carrier frequency of the laser light. Now, to work as an interferometer, the two arm lengths must be nearly the same. Let us factor out $e^{i\omega_c T_x/2}$ from both terms in Ψ. This leads to

$$\Psi = \frac{A_0}{2} e^{i(\omega_c T_x/2)} \left(e^{i(\omega_c T_x/2)} - e^{-i(\omega_c T_x/2) + i\omega_c T_y} \right). \tag{8.49}$$

Next, we also factor out $e^{i(\omega_c T_y/2)}$. This leads to

$$\Psi = \frac{A_0}{2} e^{i(\omega(T_x+T_y)/2)} \left(e^{i(\omega_c(T_x-T_y)/2)} - e^{-i(\omega_c(T_x-T_y)/2)} \right)$$

$$= i A_0 e^{i(\omega_c(T_x+T_y)/2)} \sin\left(\frac{\omega_c(T_x - T_y)}{2} \right). \tag{8.50}$$

The detected quantity is the intensity of the light, which is $|\Psi|^2 = \Psi\Psi^*$, or

$$|\Psi|^2 = A_0^2 \sin^2\left(\frac{\omega_c(T_x - T_y)}{2} \right). \tag{8.51}$$

Substituting in from Equation (8.47) for $T_x - T_y$, we obtain

$$|\Psi|^2 = A_0^2 \sin^2\left(k(L_x - L_y) + k\frac{(L_x + L_y)}{2} h_+ \cos(\omega t) \right), \tag{8.52}$$

where k is the wavenumber of the laser, and ω is the frequency of the gravitational wave.

By adjusting the difference between the arm lengths you move the intensity away from the minimum it has when $L_y = L_x$ and there is no gravitational wave signal. This is called a dark fringe, and the displacement of the intensity from the zero at a dark fringe is called an 'offset'. The intensity at this value of $L_x - L_y$ is in principle static, though you have to carefully control the mirrors to ensure that the offset is sufficiently quiet at the frequencies of interest, again around 100 Hz, to ensure that your signal is not swamped.

From the second term in the phase of the squared sine wave term in Equation (8.52) you can see that if a gravitational wave is incident on the detector, then the intensity modulates in a manner that is proportional to the gravitational wave amplitude. The amplitude of the effect is proportional to the average of the two arm lengths. This is why you build as large an interferometer as you can. This signal, a fluctuating intensity, is what was first observed by the two LIGO interferometers, operating simultaneously, a few hours before the official start of observering run 1 (O1) of the Advanced LIGO detector on 14 September 2015.

8.9.1 Shot Noise

Assuming that you can eliminate the many technical noise sources, you inevitably encounter fundamental noise due to quantum mechanics. For the first 15 years of its operation, LIGO's ultimate limiting noise floor at frequencies above about 100 Hz was the so-called standard quantum limit. Squeezed light can also be used to evade the standard quantum limit, but further discussion of squeezing is well beyond the scope of this book. The noise level you can expect from quantum mechanics assuming the standard quantum limit is treated in this section. The origins of quantum noise in LIGO can be thought of as arising from the error in measuring the relative phase of light beams from the two arms recombining at the beamsplitter.

Light beams are not in reality plane waves, but consist of coherent superpositions of quanta, which can semi-classically be considered to be wave packets. Because these packets are finite in extent, they have an inherent spread in frequency $\delta \nu$ related to the time it takes the wave packet to pass a nodal point, δt, by $\delta \nu \, \delta t \simeq 1$. Consider the thought experiment illustrated in Figure 8.5.

The laser beam is split into two components, with one component travelling down an extra section of beam path (just section from now on) of length L and acquiring an extra phase shift Φ_u. When the beams recombine, the amplitude of the summed signal can be used to determine Φ_u and hence L. A wave packet traverses the section in time $\Delta t = L/c$. The average number of packets N in the section depends on the intensity of the laser. The actual number in the section at any given time has a statistical spread σ_N due to the packets near the end overlapping with the boundaries. The size of this spread is given by $\sigma_N/N = \delta t/\Delta t$.

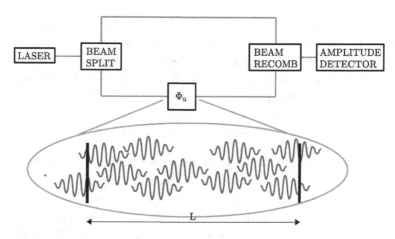

Figure 8.5 A thought experiment on measuring the difference in length between two beam paths using a coherent stream of wave packets, a simple model for a beam of laser light.

The apparatus measures the phase Φ_u picked up by the beam due to the extra path length L. Measurement of this phase will have a statistical uncertainty σ_Φ because the beam amplitude is the sum of the contributions of discrete wave packets. The phase change during the passage of the beam over a fixed point can be measured by placing a test particle at that point, which feels a force proportional to the beam amplitude. For a classical plane wave, this motion would be sinusoidal, and the phase change would be linear with increasing time. However, because the beam is made up of discrete wave packets, the motion of the particle will not be a pure sinusoid at the carrier frequency, but will instead contain a range of frequency components reflecting the sum of the effects of the discrete wave packets.

Consider the contribution of a single packet to a phase measurement. The rate of change of phase as the center of the wave packet passes over the phase detector is the laser carrier frequency. However, as the leading or lagging edge of a wave packet is passing the detector, the phase shift determined by multiplying the carrier frequency by time differs from that determined by monitoring oscillations of the test particle due to divergence of the wave packet. The rate at which these phases diverge is the frequency spread $\delta\omega$. Because all the wave packets are identical, the rate of change of phase at the center of any packet is an equally good measure of the carrier frequency. During the time between packet centers passing over the detector, the phase wanders away from linear increase with time expected for a pure sinusoid. The phase drift has an amplitude given by the frequency spread of a packet multiplied by the average time between them.

Consider placing a phase detector at one end of the section. The phase detector records amplitude as a function of time over a time interval $c\Delta t$. The resulting spread in phase σ_Φ is given by $\sigma_\Phi = \delta\omega\Delta t/N$, where $\delta\omega$ is the frequency spread in a single wave packet, and $\Delta t/N$ is the average time interval between passes of the centres of wave packets over the phase detector. The product of the spread σ_Φ in the phase shift in a length L of beam and the spread σ_N in the number of wave packets in that length is therefore

$$\sigma_N \sigma_\Phi = \frac{N\delta t}{\Delta t} \times \frac{\delta\omega\,\Delta t}{N} = 1. \tag{8.53}$$

This is the uncertainty principle as applied to laser beams in the normal coherent state of laser light. If the laser power is P, then the photon flux in the beam is $N = \lambda P/(hc)$, where λ is the laser wavelength. Since photon fluxes obey Poisson statistics, the second-to-second standard deviation in the flux is $\sigma_N = \sqrt{\lambda P/(hc)}$. From Equation (8.53) the rms of the measured phase is $\sigma_\Phi = \sqrt{hc/(\lambda P)}$. To convert this to the equivalent noise level in the gravitational wave strain, we write $\sigma_\Phi = k\sigma_L = kL\sigma_h$, where σ_L is the equivalent noise in the lengths of the interferometer arms, k is the wavenumber of the laser light, L is the average arm length, and σ_h is the noise level in the strain. This leads to a theoretical noise floor in the instrument of

$$\sigma_h = \frac{1}{2\pi L}\sqrt{\frac{hc\lambda}{P}}. \tag{8.54}$$

The most powerful continuous-beam lasers suitable for use in gravitational wave detectors have powers of order $50\,\mathrm{W}$. For $4\,\mathrm{km}$ arms at a laser wavelength of $1.06\,\mu\mathrm{m}$, the strain noise level is 2.6×10^{-21}. This is about a factor of 3 larger than the expected signal amplitude in an optimistic signal model. Clearly, an unmodified Michelson interferometer is not suitable for detecting the signal amplitudes expected from realistic astrophysical sources.

8.9.2 Resonant Cavities

Gravitational wave interferometers typically overcome the sensitivity deficit alluded at the end of Section 8.9.1 by using optical storage cavities. Here we discuss two contexts for the use of such cavities, both employed in LIGO.

Firstly, so-called arm cavities are used to increase the effective length of the interferometer arms. Extra mirrors are inserted at the near ends of the two arms, so that laser light is resonant in each arm between these near 'intermediate test mass (ITM)' mirrors and the 'end test mass (ETM)' mirrors. Simplistically, photon wave packets entering the arms undergo many round trips before leaving the arm, so that the accumulated phase shift in the photon due to any strain signal is multiplied by the number of round trips made by the wave packet. This number is of order of the finesse of the cavity, where finesse values of around 100 are achievable in practice. Since the strain noise level from Equation (8.54) is proportional to $1/L$, this enhances the signal-to-noise ratio from 1/3 to 30 for the nominal signal model assumed in Section 8.9.1.

Secondly, as alluded to in Section 8.9, the interferometer is operated at a small 'offset' from a dark fringe. Consequently, only a small fraction of the input light reaches the output photodetectors. The rest of the light must therefore exit back towards the laser. A mirror placed between the laser source and the beam splitter, called a 'recycling mirror (RM)' instead causes light between this mirror and the two arms to resonate further. This enhances the power at the beamsplitter by a further factor of the cavity finesse, which is again about 100. The laser power appears in Equation (8.54) as a denominator square root, so we can expect a further enhancement in the strain noise by a factor of 10, bring the signal-to-noise ratio up further to around 300, again for the nominal assumptions made in Section 8.9.1. This sounds like a really great signal-to-noise ratio, but bear in mind that the signals are in practice mostly smaller than those assumed here, and also the more optimistic signal models result in transients that last from a few seconds for a neutron star binary source to a small fraction of a second for two black holes. A significant signal-to-noise ratio is in practice needed to reduce the false alarm rate associated with noise sources to a manageable level.

There are several drawbacks to the use of resonators. Firstly, a successful resonator, especially a 4 km one, must be exquisitely aligned. In practice, a significant experimental effort was necessary to perfect the art of aligning such long cavities. The early development work on alignment is well described by Mavalvala (1997). Secondly, a subtle but important storage time effect occurs at higher frequencies that tends to reduce the sensitivity of the interferometers above a critical frequency. This effect is known as the 'cavity pole'.

A thorough technical discussion of the cavity pole can be found in Rakhmanov et al. (2002). We include an overview of the effect here partly because this is a critical experimental constraint on detectors in practice and partly for philosophical reasons. Right at the beginning of this book, we stated that observers in free fall making local measurements cannot detect the effects of gravity. Yet, we have shown that using mirrors which are effectively moving on geodesics we can detect interferometric signatures of gravitational waves! How is it that we are not violating the principle of equivalence here? I believe the answer is in the word local. A gravitational wave detector in fact makes a non-local measurement by studying the tidal forces on test masses well separated in space! The light travel time between such test masses constitutes a non-local observable. The masses are to be sure moving on geodesics, but the geodesics are sufficiently far apart that their relative motion can be determined. However, because light can only travel at finite speeds, effects due to time evolution of the mirror positions whilst the probe photons are between the mirrors cannot be avoided. Consider an extreme case where the time spent by a photon in the resonator may equal the period of the gravitational wave. In this case the round trip distance travelled by the photon is the same as that in the absence of the signal, since the mirror separation both decreases and then decreases over a full gravitational wave period. This is an extreme case, so let us see what happens for a more general signal.

The following argument follows that in Rakhmanov et al. (2002). Figure 8.6 shows a diagram of a resonator driven by an external light source together with a model for signal pathways into the resonator and circulating in both directions between the mirrors.

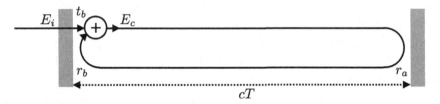

Figure 8.6 An illustration of the circulating fields between two parallel mirrors in a cavity driven by an external input beam.

We seek a steady-state solution in which the circulating fields are all oscillating at the same frequency ω and have settled to constant amplitudes. Consistency around the loop requires the following relationship between the electric fields in the incident beam and the left and right propagating circulating fields:

$$E_c = E_i t_b + r_a r_b e^{2i\omega T} E_c$$

$$\frac{E_c}{E_i} = \frac{t_b}{1 - r_a r_b e^{2i\omega T}}. \tag{8.55}$$

Now the circulating field E_c is measured, usually via a pick-off extracting the beam reflected back through the end mirrors. The phase of this beam is used to control the positions of the end mirrors such that the fields are resonant between the mirrors. The resonance condition here is that $2\omega T$ is a multiple of 2π, so that the phase factor is unity. When the reflection coefficients r_a and r_b are close to 1, the magnitude of E_c exceeds that of E_t, and power builds up between the mirrors of the resonator. Disturbances of different varieties can affect the resonator. First, the length of the cavity can change, hopefully due to a gravitational wave. This introduces a phase shift kL_h, where L_h is the length shift due to gravitational waves. In practice, L_h will oscillate. Another variety of disturbance is a change in the frequency of the light. Call the frequency disturbance ω_D. In practice, ω_D will often itself represent an oscillating component to the frequency. The overall phase factor is now $\Phi = 2\omega T + 2kL_h + 2\omega_D T$. The quiescent phase shift $2\omega T$ is set to zero by the feedback controller. Up to constants, you can see that the cavity signal E_c depends in the same way on an oscillating frequency disturbance and an oscillating cavity length disturbance. The transfer function is the ratio of E_c/E_i in the case where $\omega = \omega_D$ to the same ratio in the case where there is a non-zero ω_D. Here the transfer function is written in terms of the frequency ω_D or in terms of the variable $s = i\omega_D$:

$$H(s) = H(i\omega_D) = \frac{1 - r_a r_b}{1 - r_a r_b e^{-2i\omega_D T}} = \frac{1 - r_a r_b}{1 - r_a r_b e^{-2sT}}. \tag{8.56}$$

The frequency response of the cavity is determined by the analytic structure of $H(s)$. This function has an infinite number of poles, which are the values of s where it becomes infinite. These poles are at complex values of the frequency, where the real part of the complex frequency (which is the imaginary part of s) represents the cavity resonant frequency, and the imaginary part of the complex frequency (which is the real part of s) represents losses due to light leakage through the imperfectly reflecting mirrors, or other loss mechanisms.

These poles are the values of s where $r_a r_b \exp(-2sT) = 1$. Let us find them. There are an infinite number because we can add any multiple of $2i\pi$ to the exponent and generate another pole. Solving for the poles,

$$e^{-2sT+2ni\pi} = \frac{1}{r_a r_b}$$

$$e^{2sT-2ni\pi} = r_a r_b$$

$$2sT = \ln(r_a r_b) + 2ni\pi$$

$$s = -\frac{|\ln(r_a r_b)|}{2T} + in\frac{\pi}{T}. \tag{8.57}$$

The lowest frequency pole, often called the cavity pole because it affects the sensitivity of the detector, is at $s = -1/\tau$, where τ is the storage time of the cavity,

$$\tau = \frac{2T}{|\ln(r_1 r_2)|}. \tag{8.58}$$

In Figure 8.7, I have plotted $H(f)$, where $f = \omega/(2\pi)$. I have used $r_a = 1$ and $r_b = 0.985$ and a cavity length of $4\,\mathrm{km}$. The corresponding storage time is $\tau = 1.76\,\mathrm{ms}$, and the frequency of the cavity pole is

$$f_p = \frac{1}{2\pi \tau} = \frac{|\ln(r_1 r_2)|}{4\pi T}, \tag{8.59}$$

which is 90 Hz with these parameters. Above this frequency, the sensitivity of the cavity to length disturbances drops proportional to frequency. Of course, theoretically, the sensitivity recovers again at higher frequencies where there are further resonances, but in practice these frequencies at many tens of kilohertz, and instrumenting them at high sensitivity would require a significantly more expensive

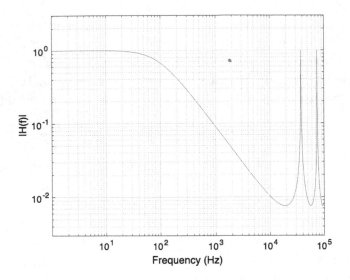

Figure 8.7 The frequency response of a cavity to perturbations on its length. The cavity is locked to the laser feeding it in the sense that a control system ensures that the nominal drive wavelength is resonant between the mirrors. The cavity pole is at about 90 Hz, above which the response to incoming signals drops off proportional to $1/f$ until above 10 kHz.

readout that, given the low likelihood of sources at these frequencies, has not been implemented in practice. The effect of the cavity pole is often represented as an increase in the level of the shot noise floor proportional to frequency above the cavity pole frequency. Below the pole frequency, the shot noise floor is flat, i.e., independent of frequency. The cavity pole effect is also manifest in designs for space-based gravitational wave detectors. Though in LISA there are no resonators because you lose most of the light between the satellites to beam divergence, the equivalent technology, which is to set up master–slave laser pairs between satellites, is susceptible to the same storage time and delay effects as the LIGO resonators, but at far longer storage times because of the vastly greater distances between the mirrors. This is why the sensitive frequency band of LISA is so much lower in frequency than that in ground-based interferometers, although the lower frequencies promise a plethora of sources, particularly white dwarf binaries, which should be abundant in the millihertz range.

8.10 Summary of Gravitational Waves

In this chapter, we have shown how linearisation of the vacuum Einstein equations in a background flat space-time leads naturally to gravitational waves. The waves are transverse to a class of observers for whom test particles in free fall do not move in the presence of gravitational waves, but the light travel times between separated freely falling bodies oscillate as the waves pass through. We identified two polarisations, orthogonal in the sense that the forces on test masses in the lab due to gravitational waves are orthogonal to each other. Detectors for gravitational waves that use laser interferometry were described in very simple terms. The Michelson interferometer technique succeeds in separating the signal, which makes the arm lengths oscillate in antiphase, affecting the so-called 'differential length', from many classes of background, such as fluctuations in the laser frequency, which in contrast affect the average arm length. The difference between the lengths of the arms is determined using laser interferometry between the reflecting surfaces of the suspended masses. This leads to the fundamental noise floor for gravitational wave interferometers, arising from the nature of laser light. An analysis of quantum fluctuations in laser beams led to an estimate of the noise floor of such instruments, and it is clear that simple Michelson interferometers will not achieve the required sensitivity. We then discussed the use of a recycling cavity to increase light intensity at the beam splitter, and arm cavities, which improve the phase sensitivity of the instrument. The latter enhancement is at the price of introducing the so-called cavity pole, whereby the sensitivity of the instrument starts to drop above a frequency corresponding to the reciprocal of the storage time of laser light in the cavities. However, with suitably highly reflecting mirrors, this frequency can be set at around 100 Hz, so that the interferometer is rendered sensitive in a frequency range where, fortuitously, an unexpectedly rich seam of sources has been struck, particularly with regard to black hole binaries at tens of solar masses.

The direct detection and subsequent study of gravitational waves from astrophysical sources may well turn out to be one of the most important scientific achievements in my lifetime, opening as it does a completely new window on the observable Universe.

8.11 Problems

8.1 Suppose I start with the LIGO interferometers and double the masses of the suspended mirrors. Does this affect the amplitude of the motion from gravitational waves? If so, what is the effect? If not, why not?

8.2 By analysing the motion of a simple pendulum in the small oscillation limit driven by a force perpendicular to the suspension wire, show that the response of the pendulum to an oscillating driving force well above the natural resonant frequency of the pendulum is the same as would be expected for an entirely free test mass floating in a gravitational field-free region.

8.3 Consider the expressions for the lab coordinates of the test masses from Equations (8.32) in the light of the general expressions for the gauge transformations (8.22). Work out expressions for the transformation components $\varepsilon^\alpha(x)$. Now figure out what components of the metric are non-zero in the lab frame. Hint: components other than the those in the XY plane become non-zero.

8.4 Gravitational wave interferometers are controlled by means of modulation sidebands added to the laser carrier. For example, an electrooptic modulator (EOM) modifies a plane wave whose time variation can be represented by $e^{i\omega_c t}$ to one whose time variation in the same representation is $e^{i\omega_c t + A_m \sin(\omega_m t)}$, where A_m is called the modulation depth, ω_c is the carrier frequency of the laser, and $\omega_m \ll \omega_c$ is the modulation frequency. Using the Laurent series

$$e^{(x/2)(t-1/t)} = \sum_{n=-\infty}^{+\infty} J_n(x)t^n,$$

where $J_n(x)$ are the Bessel functions of the first kind, show that the resultant beam contains a comb of frequency components centered at the carrier frequency, differing from the carrier frequency by integer multiples of the modulation frequency, with amplitudes that decrease with increasing order $|n|$. These are known as modulation sidebands. The technique is utilised to stabilise almost all the degrees of freedom of suspended interferometer optics.

8.5 In the quasi-static approximation, the metric signature of a space-time in the presence of a $+$ polarised gravitational wave can be written

$$ds^2 = -c^2\,dt^2 + (1+h)\,dx^2 + (1-h)\,dy^2 + dz^2,$$

where (cd, x, y, z) are the four coordinates of points in space-time, and h is the gravitational wave strain. The displacements of test particles under the influence of $+$ and $\times$ polarised gravitational waves are shown in Figure 8.8. In the absence of the gravitational wave, the test particles are each a distance L from the origin.

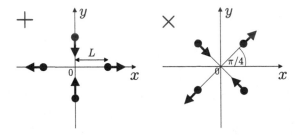

Figure 8.8 Forces on test masses under the influence of $+$ and $\times$ polarised gravitational waves.

(a) What is the magnitude of the physical displacement of the test particles from their equilibrium positions due to the gravitational wave?

(b) Draw a diagram showing x' and y' axes overlaid on the x and y axes in the left-hand figure for $+$ polarised gravitational waves, such that the test mass movements appear in your primed coordinates to be $\times$ polarised.

(c) Write the basis vectors in the primed coordinate system in terms of $\vec{e}_x$ and $\vec{e}_y$.

(d) Give the metric coefficients $g_{x'x'}$, $g_{x'y'}$, and $g_{y'y'}$.

(e) Derive expressions for the coordinates dx' and dy' of the displacement of a test particle in the primed coordinate system in terms of the displacements dx and dy of the same test particle in the unprimed coordinate system.

(f) Show that the vector displacement of the test mass positioned on the positive x axis is invariant under the transformation from unprimed to primed coordinates.

8.6 The metric signature of space-time during the passage of a $+$ polarised gravitational wave of strain amplitude h travelling in the direction of increasing z is

$$ds^2 = -c^2\,d\tau^2 = -c^2\,dt^2 + dx^2\big[1 + h\cos(kz - \omega t)\big]$$
$$+ dy^2\big[1 - h\cos(kz - \omega t)\big] + dz^2.$$

(a) Write down the Lagrangian using τ along the geodesic as the parameter.

(b) Identify two constants of the motion.

(c) Show that the Euler–Lagrange equations yield the following differential equation for the time component:

$$c^2\frac{d^2t}{d\tau^2} + \frac{\omega h}{2}\frac{A^2\sin(kz - \omega t)}{[1 + h\cos(kz - \omega t)]^2} - \frac{\omega h}{2}\frac{B^2\sin(kz - \omega t)}{[1 - h\cos(kz - \omega t)]^2} = 0,$$

where A and B are constants.

(d) Obtain an expression for $dt/d\tau$, the zeroth component of the four-velocity of a particle in free fall in the gravitational wave field. An additional integration constant will be required.

8.7 The metric in the presence of a gravitational wave incident in the z direction has the components

$$(g_{\mu\nu}) = \begin{pmatrix} -1 & 0 & 0 & 0 \\ 0 & 1 + h_+\cos(kz - \omega t) & h_\times\cos(kz - \omega t) & 0 \\ 0 & h_\times\cos(kz - \omega t) & 1 - h_+\cos(kz - \omega t) & 0 \\ 0 & 0 & 0 & 1 \end{pmatrix}.$$

(a) Show for $h_+, h_\times \ll 1$ that

$$
(g^{\mu\nu}) = \begin{pmatrix}
-1 & 0 & 0 & 0 \\
0 & 1 - h_+ \cos(kz - \omega t) & -h_\times \cos(kz - \omega t) & 0 \\
0 & -h_\times \cos(kz - \omega t) & 1 + h_+ \cos(kz - \omega t) & 0 \\
0 & 0 & 0 & 1
\end{pmatrix}.
$$

(b) Using the direct formula for the Christoffel symbols $\Gamma^\alpha_{\mu\nu}$, or otherwise, show for $h_+, h_\times \ll 1$ that

$$
\Gamma^y_{zx} = -\frac{h_\times k}{2} \sin(kz - \omega t).
$$

(c) The formula for the components A^μ of a vector being parallel transported in the x^σ direction is

$$
\frac{\partial A^\mu}{\partial x^\sigma} = -\Gamma^\mu_{\sigma\lambda} A^\lambda.
$$

Show that a vector $\vec{A}$ initially aligned parallel to the z direction will develop a non-zero y component as it is parallel transported in the direction of increasing x. Assume that $\sin(kx - \omega t)$ is constant and non-zero for the duration of the parallel transport and that $h_\times \neq 0$.

8.8 Look again at Problem 8.1. There is another reason for wanting to make the test masses heavier. Consider: a second aspect of laser beams consisting of discrete packets (photons) is that this gives rise to a noise term having to do with the interaction of laser light, particularly at high power, with suspended test masses. Identify and explain the physical effect and explain why making the test masses heavier would help to mitigate it.

9

A Guide to Further Reading

9.1 Epilogue

In this book, I have tried to deliver a self-contained introductory course in general relativity manageable by advanced and engaged undergraduates, but also useful to more advanced students. Of course, the subject is far larger and richer than the material presented here. This final chapter has three aims. Firstly, in Section 9.2, I will give some references for further study of general relativity. If you finish this course and want to keep going, what books would I recommend that you read? This is the question I want to answer here. Secondly, in Section 9.3, I will survey some of the other approaches to tensors and differential geometry. Books and online resources use many different approaches, and the student who attempts to read around the subject often becomes confused. Thirdly, in Section 9.4, I will point the student in the directions of some active research. General relativity branches out in many directions. I also give some other examples of fields in which tensor analysis is useful. The references are not intended to be a comprehensive list; rather they are a starting point for further reading. The level of difficulty of the references varies wildly, but I hope that there is something there that will pique your interest.

9.2 Books for Ongoing Study

In this book, we have restricted ourselves, firstly, to learning about tensors and, secondly, to the study of three specific problems – the Schwarzschild solution for non-spinning black holes, The Friedmann–Robertson–Walker metric of the so-called 'standard' cosmology, and the metric in the presence of propagating gravitational waves. If you want to carry on studying general relativity, then you will need to go to more advanced books. My favourite is Jim Hartle's book *Gravity, An Introduction to Einstein's General Relativity* (Hartle 2021). Hartle goes much further than I do in my book, and he has allowed himself much more space. Two topics that

I would study next are rotating black holes, which he covers in Chapter 15, and gravitational wave emission, which he covers in Chapter 23. After all, what is the point of learning about gravitational waves if you do not have any understanding of how they are created by sources? Hartle also maintains a web site with software for calculating the Riemann, Ricci, and Einstein tensors from the metric coefficients. I have not used this software extensively myself, but I can see that such tools are potentially a great saver of time and at the very least terribly useful for checking your working. I would also recommend learning more about cosmology. Knowing as you do where the Friedmann–Robertson–Walker metric comes from, you owe it to yourself to work out its consequences. In particular, I would recommend reading Ryden's book on cosmology (Ryden 2003). It is beautifully written and will help you understand the Universe. What could be better than that? Topics in cosmology that I think are particularly useful are the observable quantities such as luminosity and angular diameter distance. These are very well explained in Ryden's book.

9.3 Other Conventions in Tensor Analysis

9.3.1 The Metric Coefficients

In this book, I have adopted what is commonly referred to as the 'East Coast' convention for the coefficients of the Minkowski metric, whereby $(\eta_{\mu\nu}) = \mathrm{diag}(-1, +1, +1, +1)$. This convention is in line with that in other general relativity texts, notably Hartle (2021), Misner et al. (1973), Schutz (1985), and Weinberg (1972). The alternative 'West Coast' convention, where $(\eta_{\mu\nu}) = \mathrm{diag}(+1, -1, -1, -1)$ is adopted in other sources, notably most of the textbooks on theoretical particle physics and quantum field theory. Whichever convention you decide you like, be absolutely sure which one is used in any source you consult and do not mix results derived from sources that use alternative conventions without appropriate conversions.

9.3.2 Covariant, Contravariant, and Mixed Tensors

In the previous chapters, I have stuck fairly closely to a very conservative approach in which tensors are defined in terms of the transformation properties of their components. I have done this because I have found by experience that undergraduates approaching the subject for the first time find all the more modern approaches less easy to understand and learn. I myself learned the approach I teach from quite old, but still very useful, relevant, and authoritative texts in classical mathematical physics, notably Goldstein's book on classical mechanics (Goldstein 1980), and Panofsky and Phillips's book on classical electricity and magnetism (Panofsky and

Phillips 2012). I did not, however, make use of some terminology present in these texts, so I introduce this terminology now. These terms are covariant, contravariant, and mixed with reference to the components of tensors. A first-rank contravariant tensor is a tensor of rank $^1_0$, so its components have the following transformation law:

$$A'^\nu = \frac{\partial x'^\nu}{\partial x^\alpha} A^\alpha. \tag{9.1}$$

Note that here I have followed both of the aforementioned books and the common practice at the time in labelling the components themselves with the prime symbols rather than the indices with the primes as I did throughout the body of the text. The early texts in physics that developed tensor analysis tended not to mention the difference between the tensor itself and its components, so it therefore made more sense to attach the primes to the symbols for the components, and there were also fewer primes. Similarly, a first-rank covariant tensor is a tensor of rank $^0_1$ whose components have the transformation law

$$B'_\nu = \frac{\partial x^\alpha}{\partial x'^\nu} B_\alpha. \tag{9.2}$$

Higher-rank tensors are called contravariant and covariant if the indices on their components are all in the upstairs and downstairs positions, respectively. A tensor having components with a combination of upstairs and downstairs indices is called mixed. So $R^\phi{}_{\lambda\beta\alpha}$ are components of a mixed fourth-rank tensor in this terminology.

There is also a tendency to abbreviate the partial derivatives appearing in the transformations on covariant and contravariant tensors by writing

$$\Lambda^{\nu'}_\alpha = \frac{\partial x'^\nu}{\partial x^\alpha}, \tag{9.3}$$

so that in this notation the transformation laws for contravariant and covariant tensors are written $A'^\nu = \Lambda^{\nu'}_\mu A^\mu$ and $B'_\nu = \Lambda^\alpha_{\nu'} B_\alpha$. The choice of Λ probably originates in the use of this symbol to denote a Lorentz boost in special relativity, but here it is understood to denote a shorthand for the partial derivatives between coordinates in two general coordinate systems that you are transforming between. I have not used this notation as I do not see how it improves on just writing out the partial derivatives, but you will find it in many books, notably Schutz (1985).

9.3.3 Index-Free and Abstract-Index Notation

Throughout the book, I have tried always to say 'components of a tensor of rank'... when referring to a symbol with indices. You can also manipulate the tensors themselves rather than the components in some coordinate system. How does this work?

We have met one example. I taught you to think of a vector $\vec{A}$ as an arrow in space or space-time. Thinking purely in terms of space for a minute, as we move to different coordinate systems by rotating the axes, we look at the vector from different aspects, and it does not change the vector itself. What changes are the components of the vector and the basis vectors of the coordinate system? We got used to writing things like

$$\vec{A} = A^{\nu}\vec{e}_{\nu}. \tag{9.4}$$

The symbol $\vec{A}$ represents the tensor itself. In some sense, it is satisfying to deal in the tensors themselves. The dot product of two vectors can be written

$$\vec{A} \cdot \vec{B} = \frac{1}{4}(|\vec{A} + \vec{B}|^2 - |\vec{A} - \vec{B}|^2). \tag{9.5}$$

Furthermore, some fundamental equations can be written without reference to coordinate systems at all, notably Einstein's equations, which can be written

$$\mathbf{G} = \frac{8\pi G}{c^4}\mathbf{T} - \Lambda\mathbf{g}, \tag{9.6}$$

because the tensors $\mathbf{G}$, $\mathbf{g}$, and $\mathbf{T}$ all have the same rank. However, what is the definition of an object like $\mathbf{G}$, which we know has two index components? Here we need somehow to take the product of the two basis vectors, so we write

$$\mathbf{G} = G^{\alpha\beta}\vec{e}_{\alpha} \otimes \vec{e}_{\beta}. \tag{9.7}$$

You might wonder how to write tensors that have components with downstairs indices. The key to this is in Section 2.9.2, where we first learned that there are two representations of the components of any vector, A^{μ} and A_{μ}. They both contain the same information, but they have different transformation properties. When you contract the two, writing $A^{\mu}A_{\mu}$, the result is an invariant, a number independent of the choice of coordinates. Mathematically, we talk about A^{μ} and A_{μ} being 'dual' to each other, and the same language carries over to the components of tensors of higher rank. We have not however met the duals of the vectors themselves, so we do not know what the dual of $\vec{A}$ is, and we also do not know what the dual of $\vec{e}_{\alpha}$ is. We solve both these problems at once by defining one-forms. Symbolically, a one-form is denoted by a tilde above the quantity. The one form dual to $\vec{e}_{\alpha}$ is $\tilde{\omega}^{\alpha}$. The one form dual to $\vec{A}$ is $\tilde{A}$. Mathematically, a one-form is a function that takes as its input a single vector and produces as its output a single invariant number. The foundation of this is the basis vectors and their dual one forms, where we have

$$\tilde{\omega}^{\mu}(\vec{e}_{\nu}) = \delta^{\mu}_{\nu}. \tag{9.8}$$

Any one-form can be written as a superposition of basis one-forms, just as any vector can be written a superposition of basis vectors,

$$\tilde{A} = A_\alpha \tilde{\omega}^\alpha. \tag{9.9}$$

Using this, we can see that applying a one-form to the corresponding vector is equivalent to contracting the components of the vector and the components of its dual:

$$\tilde{A}(\vec{A}) = A_\alpha A^\beta \tilde{\omega}^\alpha (\vec{e}_\beta) = A_\alpha A^\beta \delta^\alpha_\beta = A_\alpha A^\alpha = |\vec{A}|^2. \tag{9.10}$$

An important subtlety is that the one-forms $\tilde{\omega}^\alpha$ are not obtained by using the inverse metric to raise the index on the corresponding basis vectors $\vec{e}_\alpha$, so $\tilde{\omega}^\beta \neq g^{\beta\alpha} \vec{e}_\alpha$. This is because $\vec{e}_\alpha$ are not the components of a vector, the symbol instead represents one of a set of vectors obtained by setting the index α to one of four values. Taken as a set, they do not form the components of a tensor of rank $^0_1$; rather each member of the set is itself a vector. It is a subtle but important difference. Since we have one-forms acting as functions that operate on a single vector to yield an invariant number, can we also have two-forms that operate on two vectors and yield a single number? Yes, we can, and the prototypical example is the metric that takes two vectors as its input and returns their dot product, an invariant:

$$\tilde{g}(\vec{A}, \vec{B}) = \vec{A} \cdot \vec{B}. \tag{9.11}$$

You can write $\tilde{g}$ as a superposition of the direct products of basis one-forms with coefficients given by the tensor components:

$$\tilde{g} = g_{\mu\nu}\, \tilde{\omega}^\mu \otimes \tilde{\omega}^\nu. \tag{9.12}$$

In this notation, the action of $\tilde{g}$ on the vectors is that the first basis one-form takes as its argument the basis vector from the left-hand slot in the function argument, and the second basis one-form takes the basis vector from the right-hand slot. So,

$$\begin{aligned}
\tilde{g}(\vec{A}, \vec{B}) &= \tilde{g}\left(A^\alpha \vec{e}_\alpha, A^\beta \vec{e}_\beta\right) \\
&= A^\alpha A^\beta \tilde{g}(\vec{e}_\alpha, \vec{e}_\beta) \\
&= g_{\mu\nu}\, A^\alpha B^\beta \tilde{\omega}^\mu(\vec{e}_\alpha) \tilde{\omega}^\nu(\vec{e}_\beta) \\
&= g_{\mu\nu}\, A^\alpha A^\beta \delta^\mu_\alpha \delta^\nu_\beta = g_{\mu\nu} A^\mu B^\nu.
\end{aligned} \tag{9.13}$$

In the case of the metric, because $g_{\mu\nu}$ is symmetric, interchanging the $\vec{A}$ and the $\vec{B}$ arguments of $\tilde{g}$ does not change the answer. However, were a two-form to represent a function that is not symmetric, the order in which the vectors were fit into the two slots would affect the result.

In general, we can extend this index-free picture of tensors by defining a tensor of rank ^n_m as a function that takes as its arguments n one-forms and m vectors and produces an invariant. This definition is often nowadays preferred to the more old-fashioned definition in terms of the transformation properties of the components, I suspect because the definition is then a natural extension of the mathematical definition of a vector.

In my experience, however, students find this approach hard, which is why I have not adopted it. Furthermore, as soon as you actually want to calculate anything, a necessary requirement particularly for those of us who want to do experiments, is to pick a coordinate system, usually one where the basis vectors are dictated by laboratory conditions. We have encountered this reality in Chapters 6–8, because naturally having worked out the theory of the Schwarzschild solution, standard cosmology, and gravitational waves, we wish to do experiments to check that it is all correct! An entertaining summary of the situation and a rather nice pictorial representation of the modern definition of a tensor given in the previous paragraph are provided by the 'popular science' book (in inverted commas because it is a brave lay-person who embarks on reading this book!) by Penrose, *The Road to Reality* (Penrose 2005).

A final twist on index-free notation is the so-called abstract index notation. In this notation, which is used in particular in the advanced book by Wald (1984), we label the tensors themselves, rather than their components, with latin indices. So A_a is a one-form, g_{ab} is a two-form, and A^c is a vector. The understanding is that the indices are there to indicate the tensor nature of the object, as opposed to referring to the components of that object in some particular coordinate system. So we could bridge the gap between abstract index notation by writing

$$g_{ab} = \tilde{g} = g_{\mu\nu}\,\tilde{\omega}^\mu \otimes \tilde{\omega}^\nu. \tag{9.14}$$

This notation leads to equations that strongly resemble the equations in components that I have used in this book. Personally, I do not see the practical difference between writing the component equations themselves and using indices in all the same places as signposts for the transformation laws, though there is certainly value in appreciating the distinction between a tensor and its components. I am therefore equally content with abstract index notation and component notation.

9.3.4 Basis Vectors Re-invented as Operators

A further set of conventions and notations surrounds the differential geometry heritage of general relativity, and this notation has also been used in the literature, notably again by Misner, Thorne, and Wheeler (1973) and also in the textbook by Ryder (2009) and on many sites related to general relativity on the internet. Again, the results of application of this notation are identical to the results of applying the component formalism, but the mode of thinking about the meaning of the mathematical symbols is different. We start with one of the most practically useful equations from Chapter 2, that for deducing expressions for the basis vectors in some coordinate systems. This is first written down in Equation (2.8) but then used many times over in the ensuing examples and problems:

$$\vec{e}_\mu = \frac{\partial \vec{r}}{\partial x^\mu}. \qquad (9.15)$$

The vector $\vec{r}$ was introduced in Cartesian coordinates and flat spaces as a vector from the chosen origin to any arbitrary point in the space. The tensor formalism led to transformation laws that allowed us to take expressions in terms of tensor components derived in 'flat' spaces and use them in arbitrarily curved spaces and in curvilinear coordinates that we may choose to use for convenience of calculation even in flat spaces.

If we start to think about what the symbols might mean in curved spaces, then we might start to worry. Imagine we are on the surface of an apple. We choose an origin at some place on the surface of the apple, and we try to define a vector $\vec{r}$ from that point to some other point on the surface. This vector leaves the surface and travels through the flesh of the apple, so there is an implicit assumption that the curved surface is embedded in a flat higher-dimensional space; otherwise, it becomes basically impossible to define $\vec{r}$ meaningfully. This makes the above definition of a basis vector somewhat unsatisfactory. So, we could define the basis vector instead as an operator simply by removing the offending $\vec{r}$. Because I do not want this new object to get confused with the previous definition, I have given it a different symbol, using the germanic Fractur script to denote the new operator, distinguishing it from $\vec{e}_\mu$ as defined in Equation (2.8),

$$\vec{\mathfrak{e}}_\mu = \frac{\partial}{\partial x^\mu}. \qquad (9.16)$$

Suppose we have some scalar function of position, say $f(x)$. We also have a pathway in our space; perhaps, it is a geodesic, along which a parameter τ tells us our position. If we move a small distance $d\tau$ along this pathway, then the function f changes by an amount

$$df = \frac{df}{d\tau} d\tau. \qquad (9.17)$$

In practice, to figure out the result df, or more commonly to get the large change in f due to moving a large distance along the pathway, we need to specify the shape of the path in terms of a coordinate system. So we play the usual trick by invoking the total derivative

$$df = \frac{\partial f}{\partial x^\mu} \frac{dx^\mu}{d\tau} d\tau. \qquad (9.18)$$

However, if we know the shape of the geodesic in the space, $x^\mu(\tau)$, then we can express df in terms of the change in coordinates dx^μ that keeps you on the geodesic as you move:

$$df = \frac{\partial f}{\partial x^\mu}\, dx^\mu. \tag{9.19}$$

Inserting our newly defined basis vector operator, we can write this as

$$df = \vec{e}_\mu(f)\, dx^\mu. \tag{9.20}$$

Geometrically, the components dx^μ define the direction in which the geodesic is pointing in the space. You can think of an arrow taking off in a direction tangent to the geodesic at the point in question. What about $\vec{e}_\mu$? you can think of this geometrically as defining a set of parallel planes whose normal is in the direction defined by a line of constant x^μ. The spacing between the planes varies inversely as the magnitude of f. As f gets smaller, the space between the planes gets larger. A number is generated by looking at how many planes are sliced through by the arrow representing dx^μ. Conceptually, this avoids the schism of having to imagine the quotient of two numbers both tending to zero having a meaningful value. Instead, we have the abstract geometrical concept of a tangent vector to the geodesic and a set of planes overlaying that tangent space. To determine the change df in some field, we can scale the length of the arrow dx^μ by some factor, say a, as long as we divide by this same scale factor elsewhere in the expression:

$$df = \vec{e}_\mu(f)\, dx^\mu = \vec{e}_\mu\left(\frac{f}{a}\right) d(ax^\mu). \tag{9.21}$$

Because the function operated on by $\vec{e}_\mu$ has become smaller, conceptually the planes move further apart, but the tangent arrow is now correspondingly longer, so the number of planes cut through does not change. This construction allows us to get away from always having to think about the limit of small dy and dx in a derivative dy/dx. The concept is replaced by the idea of a tangent space overlaid with planes. The tangent space only touches the geodesic at one point, but that is fine because this is all we need it to do to calculate the change in f. All that remains is to figure out what corresponds to the one-forms in the index-free picture. The basis vectors became a set of operators on some field (any field for that matter – it does not have to be a scalar field), but to generate the output number df, we need a function of that set of operators, which geometrically corresponds to the direction in which the geodesic is going. This is supplied by dx^μ, so we define

$$\tilde{w}^\mu = dx^\mu. \tag{9.22}$$

The one-form maps the vector $\vec{e}_\mu(f)$ onto the scalar, df, so that

$$\tilde{w}^\mu\left(\vec{e}_\mu(f)\right) = \vec{e}_\mu(f)\tilde{w}^\mu = \frac{\partial f}{\partial x^\mu}\,dx^\mu = df. \tag{9.23}$$

The one-forms are dual to the basis vectors in the sense that

$$\vec{e}_\mu\left(\tilde{w}^\nu\right) = \frac{\partial}{\partial x^\mu}\,dx^\nu = \delta^\nu_\mu. \tag{9.24}$$

My personal recommendation is that whenever you encounter a new source of help or inspiration on general relativity, you root around in that source and discover, firstly, what convention they use for the sign of the metric coefficients and, secondly, how they define a basis vector. Otherwise, you may end up mixing results obtained in the context of the different frameworks with, in my experience, fairly disastrous results. Those of you who are interested in mathematical differential geometry can find huge amount of information in the literature. Again a useful reference is Misner, Thorne, and Wheeler (1973). A very authoritative but formidable source is the multi-volume reference by Spivak (1999). A shorter but quite terse and compact set of notes is the notes by Eguchi, Gilkey, and Hanson (1980). As well as underpinning general relativity, the subject has myriad other applications. It is probably a lifetime's work to achieve a broad understanding of this fascinating material.

9.4 Where from Here?

General relativity and the tensor formalism that enables it have an enormous number of research applications. Firstly, there is the programme of experimental work testing the predictions of the theory. Important experiments making precision measurements of gravity at all distance scales. In recent years torsion balance experiments (Adelberger et al. 2009), lunar ranging experiments (Merkowitz 2010), Shapiro time delay measurements, and satellite measurements of frame dragging and the geodetic effect with Gravity Probe B (Everitt et al. 2011) have added to the historic tests using the bending of light from the Sun during solar eclipses and the precession of the perihelion of Mercury. The paper by (Will 2014) provides a comprehensive review. The most significant experiments in gravitation in recent years have been aimed at direct detection of gravitational waves using ground-based laser interferometers. The LIGO (Aasi et al. 2015), Virgo (Accadia et al. 2012), GEO600 (Willke et al. 2002), TAMA-300 (Tsubono 1995), and KAGRA (Akutsu et al. 2019) experiments were enabled by earlier prototype instruments, which themselves pioneered and tested many of the enabling technologies including the 40 m prototype at Caltech (Driggers 2015), the FMI Mavalvala (1997), and PNI Lantz (1999) interferometers at MIT, the 30 m prototype at Garching (Shoemaker et al. 1988), and the 10 m prototype at Glasgow (Robertson et al. 1995). The first direct detection of gravitational waves in 2015 by advanced LIGO (Abbott et al. 2016) was the crowning achievement of 45 years of experimental work by thousands of scientists, confirming Einstein's prediction of gravitational waves, which was auspiciously almost exactly 100 years old at the time of the discovery. Kip Thorne, Barry Barish, and Rainer (Rai) Weiss shared in the Nobel Prize for physics in 2017 (Weiss 2018).

Other experiments searching for gravitational waves have used lunar ranging (Merkowitz 2010), pulsar timing (Detweiler 1979), and resonant bar (Aguiar 2010) techniques. The programme was spurred on by the indirect evidence for gravitational waves cleverly unearthed by Hulse and Taylor through radio observations of the spin-down of binary pulsar PSR 1913+16 as it radiates the energy contained in its angular momentum as gravitational waves (Hulse and Taylor 1975; Taylor and Weisberg 1982), a result that also led to a Nobel Prize in 1993. Since the first direct detections by LIGO in 2015 (Abbott et al. 2016) and the detection of the first gravitational wave signature of a binary neutron star collision in 2017 (Abbott et al. 2017), as well as order of 100 further discoveries by the Advanced LIGO, Virgo, and KAGRA network (Abbott et al. 2021), gravitational wave interferometry is established as an observational branch of astrophysics and astronomy, with many more discoveries promising to provide important data for astrophysics and cosmology well in to the future. Exciting future experiments include LISA to probe low-frequency gravitational waves from a network of satellites (Vitale 2014) and on the ground, LIGO A+ (Barsotti et al. 2018), Voyager (Adhikari et al. 2020), Cosmic Explorer (Dwyer et al. 2015), and the Einstein Telescope (Punturo et al. 2010). Additional work in atom interferometry in the future might provide a further method to detect gravitational waves using coherent beams of atoms in place of lasers (Geiger 2017). Since atomic beams have far lower velocities than laser beams, atomic beam interferometers promise sensitivity to low-frequency gravitational waves in terrestrial instruments. The challenge is to achieve coherence between atomic beams over sufficiently long distances, but experimentalists are ingenious, and this challenge may be overcome.

On the theoretical side, we have examined two exact solutions and one approximate solution of Einstein's equations. There are other known exact solutions, for example, the Kerr solution for a spinning mass distribution (Hartle 2021), and beyond that, there is a whole research field of studies of exact solutions to Einstein's equations (Stephani et al. 2003). There is also much activity in theoretical work to unify the theory of gravity with the quantum field theories that describe the interactions of particles via the strong, weak, and electromagnetic forces (Howl et al. 2019). There are various clues as to where these solutions may lie, and geometrical methods are central to the theoretical work that is being done. Extensions to the standard model of particle physics such as supersymmetry provide clues as to how gravity and quantum field theory might be reconciled, though many difficulties remain, notably the lack of experimental evidence for supersymmetry (Tanabashi et al. 2018). A further area of active research is the question of information and entropy in a gravitational field. What happens to the information content of material falling in to a black hole, for example (Polchinski 2015)? Part of the solution may lie in the correspondence between conformal field theory and anti-de Sitter space, often abbreviated to the 'AdS/CFT correspondence' (Hubeny 2015). In general, the active theoretical work around quantum measurement and quantum computation is feeding a re-examination of quantum field theory. It is John Wheeler who coined the phrase 'it from bit', which implies that fundamental new insights into the structure underpinning physics may come from the study of quantum systems and how they can interact and be entangled (Moskowitz 2016).

Between pure theory and experiment there lies the field of numerical relativity. Though Einstein's equations are famously difficult to solve exactly, the use of approximations known collectively as post-Newtonian relativity can lead to predictions that are tested against astronomical observations. Indeed, the determination of the physical parameters of the sources of the LIGO and Virgo gravitational waves was done using computer simulations in the framework of post-Newtonian mechanics. A particularly fine book by Poisson and Will (2014) is a great springboard into this enormous area, through which a very large community of theoretical physicists have made inroads into the structure of compact objects and a whole range of practical questions about gravitational phenomena in the observed Universe.

Tensor formalism itself has applications well beyond gravitation. Many of the equations you will be familiar with from your courses are in fact tensor expressions. For example, the polarisation vector $\vec{P}$ in a dielectric is related to the electric field $\vec{E}$ by $\vec{P} = \varepsilon_0 \chi_e \vec{E}$, where χ_e is the dielectric susceptibility of the material. There is no reason in general why the induced polarisation should be parallel to the applied electric field, so in general χ_e is a rank 2 tensor. Similarly, when you sit on a block of rubber, your weight exerts a downward force, but the rubber material is pushed outwards sideways. So the Young modulus of the rubber is a tensor too. There are many other examples of problems where tensor analysis is bought to bear in fluid dynamics, plasma physics, materials physics, and almost all other areas! An authoritative summary of classical physics is delivered by the encyclopaedic volume by Thorne and Blandford (2017), as well as by several of the volumes from the 'theoretical minimum' series by Landau and Lifschitz, particularly the slim volume on the theory of elasticity (Landau et al. 1986). On a more mundane level, we have seen in this book several examples where tensor methods can be used to work out the form taken by vector identities in physics, so these methods achieve relevance to undergraduate physics as well.

The formalism of differential geometry and tensor analysis also finds applications in theoretical particle physics quite independent of the above-mentioned efforts to unify gravity and field theory. Modern particle physics is formulated as a gauge theory, which means that the wave functions describing particle states are symmetric under position-dependent transformations belonging to representations of groups. The analogy between the transformations that are applied to tensors on coordinate transformations and the transformations applied to particle states is almost exact. For example, the Christoffel symbols that serve to connect the transformations applied to basis vectors at neighbouring points are analogous to the vector potential, which serves to connect the gauge transformations applied to spinors describing electrons at neighbouring space-time points. Indeed, in quantum mechanics the definition of tensors is made more widely than we have in this course and encompasses objects transforming in a known set of manners by any representation of any continuous transformation group under which the mathematical functions representing any physical object we can describe theoretically. A book by Georgi (1999) is a good place to read about the transformations of quantum fields by representations of groups. Gauge theories of particle physics and the standard model are themselves the study of a lifetime. My favourite textbooks in this area are the book by Peskin and Schroeder (1995) on quantum field theory, the lectures of Coleman (2018) published posthumously, and the book by Schwartz (2014) on the standard model of particle physics. Further applications of differential geometry to particle physics extend well beyond the standard model, for example, into supersymmetry. An introductory text in this area is Wess and Bagger (1992).

The application of general relativity to cosmology is yet another area of vigorous activity. Observations of the Universe on large scales seem to validate the assumptions of homogeneity and isotropy that underpin the standard model of cosmology. However, the subject is not without its puzzles. Central amongst these are the twin enigmas of dark matter and the so-called dark energy. Given that we sometimes tell ourselves how much we have learned about the Universe, it is a little embarrassing to admit how little of the energy and mass present in our Universe we have actually understood. Many of you will be taking cosmology courses that start with the Friedmann equations derived in Chapter 7 and will encounter observational tests of the ideas laid out there. There are many good books, but the one I recommend the most is Ryden's book *Introduction to Cosmology* (Ryden 2003) as an authoritative yet friendly and clear introduction to the subject. An older, and far more advanced, but still very useful book that successfully presents both general relativity and cosmology in detail is Weinberg (1972). Along with Weinberg, other more advanced books on cosmology include Kolb and Turner (1990), Peebles (1994), and the two-volume set by Rubakov and Gorbunov (2017; 2011).

In closing this chapter and the book, I only hope that the long and perhaps some-what exhausting list of areas into which those familiar with general relativity can branch out does not intimidate students who have begun their studies with this book. The author has had a lifetime of fun and enjoyment out of delving into this material. Indeed, the difficulty in writing this book was in deciding what to include and what to leave out! Perhaps, this epilogue and the references therein will spur you on to hit the library and the internet and explore the universe of possibilities. I wish all students beginning to study this subject a lifetime of happy occupation in exploring and thinking about the Universe in which we are privileged to live.

References

Aasi, J., Abbott, B. P., Abbott, R., et al. 2015. Advanced LIGO. *Classical Quantum Gravity*, **32**(7), 074001.

Abbott, B. P., Abbott, R., Abbott, T. D., et al. 2016. Observation of gravitational waves from a binary black hole merger. *Phys. Rev. Lett.*, **116**(Feb), 061102.

Abbott, B. P., Abbott, R., Abbott, T. D., et al. 2017. GW170817: Observation of gravitational waves from a binary neutron star inspiral. *Phys. Rev. Lett.*, **119**(Oct), 161101.

Abbott, R., Abbott, T. D., Acernese, F., et al. 2021. *GWTC-3: Compact Binary Coalescences Observed by LIGO and Virgo During the Second Part of the Third Observing Run*.

Accadia, T., Acernese, F., Alshourbagy, M., et al. 2012. Virgo: a laser interferometer to detect gravitational waves. *J. Instrum.*, **7**(03), P03012.

Adelberger, E. G., Gundlach, J. H., Heckel, B. R., et al. 2009. Torsion balance experiments: a low-energy frontier of particle physics. *Prog. Part. Nucl. Phys.*, **62**(1), 102–134.

Adhikari, R. X., Arai, K., Brooks, A. F., et al. 2020. A cryogenic silicon interferometer for gravitational-wave detection. *Classical Quantum Gravity*, **37**(16), 165003.

Aguiar, Odylio Denys. 2010. Past, present and future of the resonant-mass gravitational wave detectors. *Res. Astron. Astrophys.*, **11**(1), 1–42.

Akutsu, T., Ando, M., Arai, K., et al. 2019. KAGRA: 2.5 generation interferometric gravitational wave detector. *Nature Astronomy*, **3**(01), 35–40.

Barsotti, Lisa, Collaboration, LIGO. 2018. The A+ upgrade for advanced LIGO. Pages S14–002 of: *APS April Meeting Abstracts*. Vol. 2018. American Physical Society.

Coleman, Sidney. 2018. *Lectures of Sidney Coleman on Quantum Field Theory*. World Scientific.

Detweiler, S. 1979. Pulsar timing measurements and the search for gravitational waves. *Astrophys. J.*, **234**(Dec), 1100–1104.

Driggers, J. C. 2015. *Noise Cancellation for Gravitational Wave Detectors*. Dissertation (Ph.D.), California Institute of Technology.

Dwyer, Sheila, Sigg, Daniel, Ballmer, Stefan W., Barsotti, Lisa, Mavalvala, Nergis, and Evans, Matthew. 2015. Gravitational wave detector with cosmological reach. *Phys. Rev. D*, **91**(Apr), 082001.

Eguchi, Tohru, Gilkey, Peter B., and Hanson, Andrew J. 1980. Gravitation, gauge theories and differential geometry. *Phys. Rep.*, **66**(6), 213–393.

Everitt, C. W. F., DeBra, D. B., Parkinson, B. W., et al. 2011. Gravity Probe B: final results of a space experiment to test general relativity. *Phys. Rev. Lett.*, **106**(May), 221101.

Geiger, Remi. 2017. Future gravitational wave detectors based on atom interferometry. Pages 285–313 of: *An Overview of Gravitational Waves*. World Scientific.

Georgi, H. 1999. *Lie Algebras in Particle Physics. From Isospin to Unified Theories*. CRC Press.

Goldstein, Herbert. 1980. *Classical Mechanics*. Addison-Wesley.

Gorbunov, Dmitry S., and Rubakov, Valery A. 2011. *Introduction to the Theory of the Early Universe: Cosmological Perturbations and Inflationary Theory*. World Scientific.

Hartle, James B. 2021. *Gravity: An Introduction to Einstein's General Relativity*. Cambridge University Press (previously Pearson, 2003).

Howl, Richard, Penrose, Roger, and Fuentes, Ivette. 2019. Exploring the unification of quantum theory and general relativity with a Bose–Einstein condensate. *New J. Phys.*, **21**(4), 043047.

Hubeny, Veronika E. 2015. The AdS/CFT correspondence. *Classical and Quantum Gravity*, **32**(12), 124010.

Hulse, R. A., and Taylor, J. H. 1975. Discovery of a pulsar in a binary system. *Astrophys. J. Lett.*, **195**(Jan), L51–L53.

Katz, Victor J. 1995. Ideas of calculus in Islam and India. *Math. Mag.*, **68**(3), 163–174.

Kolb, Edward W., and Turner, Michael S. 1990. *The Early Universe*. Frontiers in Physics, Vol. 69. Addison-Wesley.

Landau, L. D., Lifšic, E. M., Lifshitz, E. M., Kosevich, A. M., Sykes, J. B., Pitaevskii, L. P., and Reid, W. H. 1986. *Theory of Elasticity: Volume 7*. Course of Theoretical Physics. Elsevier Science.

Lantz, B. T. 1999. *Quantum Limited Optical Phase Detection in a High Power Suspended Interferometer*. Dissertation (Ph.D.), Massachusetts Institute of Technology.

Mavalvala, N. 1997. *Alignment Issues in Laser Interferometric Gravitational-Wave Detectors*. Dissertation (Ph.D.), Massachusetts Institute of Technology.

Merkowitz, Stephen. 2010. Tests of gravity using Lunar Laser Ranging. *Living Rev. Relativ.*, **13**(11).

Misner, Charles W., Thorne, K. S., and Wheeler, J. A. 1973. *Gravitation*. W. H. Freeman.

Moskowitz, Clara. 2016. Tangled up in spacetime. *Sci. Am.*, **316**(12), 32–37.

Newton, I., Cohen, I. Bernard, Whitman, Anne, and Budenz, Julia. 1999. *The Principia: The Authoritative Translation and Guide: Mathematical Principles of Natural Philosophy*. 1st edn. University of California Press.

Panofsky, W. K. H., and Phillips, M. 2012. *Classical Electricity and Magnetism*. 2nd edn. Dover Books on Physics. Dover Publications.

Peebles, P. J. E. 1994. *Principles of Physical Cosmology*. Princeton University Press.

Penrose, Roger. 2005. *The Road to Reality: A Complete Guide to the Laws of the Universe*. Random House.

Peskin, Michael E., and Schroeder, Daniel V. 1995. *An Introduction to Quantum Field Theory*. Addison-Wesley.

Poisson, Eric, and Will, Clifford M. 2014. *Gravity: Newtonian, Post-Newtonian, Relativistic*. Cambridge University Press.

Polchinski, Joseph. 2015. The black hole information problem. Pages 353–397 of: *Theoretical Advanced Study Institute in Elementary Particle Physics: New Frontiers in Fields and Strings*. World Scientific.

Punturo, M., Abernathy, M., Acernese, F., et al. 2010. The Einstein Telescope: a third-generation gravitational wave observatory. *Classical Quantum Gravity*, **27**(10).

Rakhmanov, M., Savage, R. L., Reitze, D. H., and Tanner, D. B. 2002. Dynamic resonance of light in Fabry–Perot cavities. *Phys. Lett. A*, **305**(5), 239–244.

Robertson, D. I., Morrison, E., Hough, J., et al. 1995. The Glasgow 10m prototype laser interferometric gravitational wave detector. *Rev. Sci. Instrum.*, **66**(9), 4447–4452.

Roy, Ranjan. 2021. Chapter 8, The calculus of Newton and Leibniz. Pages 143–164 of: *Sources in the Development of Mathematics*. 2nd edn. Vol. 1. Cambridge University Press.

Rubakov, Valery A., and Gorbunov, Dmitry S. 2017. *Introduction to the Theory of the Early Universe: Hot Big Bang Theory*. World Scientific.

Ryden, Barbara. 2003. *Introduction to Cosmology*. Addison-Wesley.

Ryder, Lewis. 2009. *Introduction to General Relativity*. Cambridge University Press.

Saulson, Peter R. 2017. *Fundamentals of Interferometric Gravitational Wave Detectors*. 2nd edn. World Scientific.

Schutz, Bernard F. 1985. *A First Course in General Relativity*. Cambridge University Press.

Schwartz, Matthew D. 2014. *Quantum Field Theory and the Standard Model*. Cambridge University Press.

Shoemaker, D., Schilling, R., Schnupp, L., et al. 1988. Noise behavior of the Garching 30-meter prototype gravitational-wave detector. *Phys. Rev. D*, **38**(Jul), 423–432.

Spivak, M. 1999. *A Comprehensive Introduction to Differential Geometry*. Publish or Perish, Incorporated.

Stephani, H., Kramer, D., MacCallum, M., et al. 2003. *Exact Solutions of Einstein's Field Equations*. 2nd edn. Cambridge Monographs on Mathematical Physics. Cambridge University Press.

Tanabashi, M., et al. (Particle Data Group). 2018. Review of particle physics. *Phys. Rev. D*, **98**(3), 030001.

Taylor, J. H., and Weisberg, J. M. 1982. A new test of general relativity – gravitational radiation and the binary pulsar PSR 1913+16. *Astrophys. J.*, **253**(Feb), 908–920.

Thorne, Kip S., and Blandford, Roger D. 2017. *Modern Classical Physics: Optics, Fluids, Plasmas, Elasticity, Relativity, and Statistical Physics*. Princeton University Press.

Tsubono, K. 1995. 300-m laser interferometer gravitational wave detector (TAMA300). In *Japan Gravitational Wave Experiments*. World Scientific.

Vitale, Stefano. 2014. Space-borne gravitational wave observatories. *Gen. Relativity Gravitation*, **46**, 1730.

Wald, Robert M. 1984. *General Relativity*. Chicago University Press.

Weinberg, S. 1972. *Gravitation and Cosmology*. John Wiley and Sons.

Weiss, Rainer. 1972. Electromagnetically coupled broadband gravitational antenna. *Q. Prog. Rep. Mass. Inst. Technol., Res. Lab. Electron.*, **105**, 54.

Weiss, Rainer. 2018. Nobel Lecture: LIGO and the discovery of gravitational waves I. *Rev. Modern Phys.*, **90**(4), 040501.

Wess, J., and Bagger, J. 1992. *Supersymmetry and Supergravity*. Princeton University Press.

Will, Clifford M. 2014. The confrontation between general relativity and experiment. *Living Rev. Relativ.*, **17**(1), 1–117.

Willke, B., Aufmuth, P., Aulbert, C., et al. 2002. The GEO 600 gravitational wave detector. *Classical and Quantum Gravity*, **19**(7).

Index

Printed in the United States
by Baker & Taylor Publisher Services

Printed in the United States
by Baker & Taylor Publisher Services